ALL THE WORLD'S ANIMALS
RODENTS

ALL THE WORLD'S ANIMALS
RODENTS

TORSTAR BOOKS
New York · Toronto

CONTRIBUTORS

RJvA Rudi J. van Aarde
University of Pretoria
Pretoria
South Africa

GA Greta Ågren
University of Stockholm
Stockholm
Sweden

RJB R. J. Berry PhD
University College
London
England

CB Claude Baudoin
Université de Franche-Comté
Besançon
France

IRB Ian R. Bishop
British Museum (Natural
History)
London
England

TMB Thomas M. Butynski
Kibale Forest Project
Fort Portal
Uganda

JAC Joe A. Chapman PhD
Utah State University
Logan, Utah
USA

GBC Gordon B. Corbet PhD
British Museum (Natural
History)
London
England

DPC David P. Cowan PhD
Ministry of Agriculture,
Fisheries and Food
Guildford
England

MJD Michael J. Delany MSc DSc
University of Bradford
Bradford
England

ACD Adrian C. Dubock PhD
Plant Protection Division
Imperial Chemical Industries
PLC
Haslemere
England

JFE John F. Eisenberg PhD
Florida State University
University of Florida
Gainesville, Florida
USA

JE James Evans
US Fish and Wildlife Service
Olympia, Washington
USA

JF Julie Feaver PhD
University of Cambridge
England

TEF Theodore E. Fleming PhD
University of Miami
Coral Gables, Florida
USA

PJG Peter J. Garson BSc DPhil
University of Newcastle upon
Tyne
Newcastle upon Tyne
England

WG Wilma George DPhil
University of Oxford
England

EH Emilio Herrera BSc
University of Oxford
England

HEH Harry E. Hodgdon
School of Forestry Resources
North Carolina State University
Raleigh, North Carolina
USA

UWH U. William Huck
Princeton University
Princeton, New Jersey
USA

JUMJ Jennifer U. M. Jarvis
University of Cape Town
Rondebosch
South Africa

TKa Takeo Kawamichi
University of Osaka
Osaka
Japan

LBK Lloyd B. Keith PhD
University of Wisconsin-
Madison
Madison, Wisconsin
USA

REK Robert E. Kenward DPhil
Institute of Terrestrial Ecology
Abbott's Ripton
England

CJK Charles J. Krebs PhD
Division of Wildlife and
Rangelands Research
CSIRO
Lyneham, A.C.T.
Australia

TEL Thomas E. Lacher, Jr BSc PhD
Huxley College of
Environmental Studies
Bellingham, Washington
USA

RAL Richard A. Lancia
School of Forestry Resources
North Carolina State University
Raleigh, North Carolina
USA

DWM David W. Macdonald MA DPhil
University of Oxford
England

KM Kathy Mackinnon MA DPhil
Bogor
Indonesia

JLP James L. Patton PhD
University of California
Berkeley, California
USA

GBR Galen B. Rathbun BA PhD
US Fish and Wildlife Service
Gainesville, Florida
USA

ES Eberhard Schneider
University of Göttingen
Göttingen
West Germany

PWS Paul W. Sherman PhD
Cornell University
Ithaca, New York
USA

ATS Andrew T. Smith BA PhD
Arizona State University
Tempe, Arizona
USA

DMS D. Michael Stoddart BSc PhD
King's College
London
England

ABT Andrew B. Taber BA
University of Oxford
England

JOW John O. Whitaker, Jr PhD
Indiana State University
Terre Haute, Indiana
USA

CAW Charles A. Woods
Florida State Museum
University of Florida
Gainesville, Florida
USA

ALL THE WORLD'S ANIMALS
RODENTS

TORSTAR BOOKS INC.
41 Madison Avenue, Suite 2900, New York, NY 10010

Project Editor: Graham Bateman
Editors: Peter Forbes, Bill MacKeith, Robert Peberdy
Art Editor: Jerry Burman
Picture Research: Linda Proud, Alison Renney
Production: Clive Sparling
Design: Chris Munday

Originally planned and produced by:
Equinox (Oxford) Ltd
Littlegate House
St Ebbe's Street
Oxford OX1 1SQ, England

Editor
Dr David Macdonald
Animal Behaviour Research Group
University of Oxford, England

Artwork Panels
Priscilla Barrett

Library of Congress Cataloging in Publication Data

Rodents.

(All the world's animals)
Bibliography: p.
Includes index.
1. Rodents. I. Series.
QL737.R6R625   1986   599.32'3   86–4335
ISBN 0–920269–78–8

ISBN 0–920269–72–9 (Series: All the World's Animals)
ISBN 0–920269–78–8 (Rodents)

On the cover: Gray squirrel
Page 1: Belding's ground squirrel
Pages 2–3: Brandt's climbing mouse
Pages 4–5: North American porcupine
Pages 6–7: Harvest mouse
Pages 8–9: Giant mole-rat

9 8 7 6 5 4 3 2 1

Printed in Belgium

CONTENTS

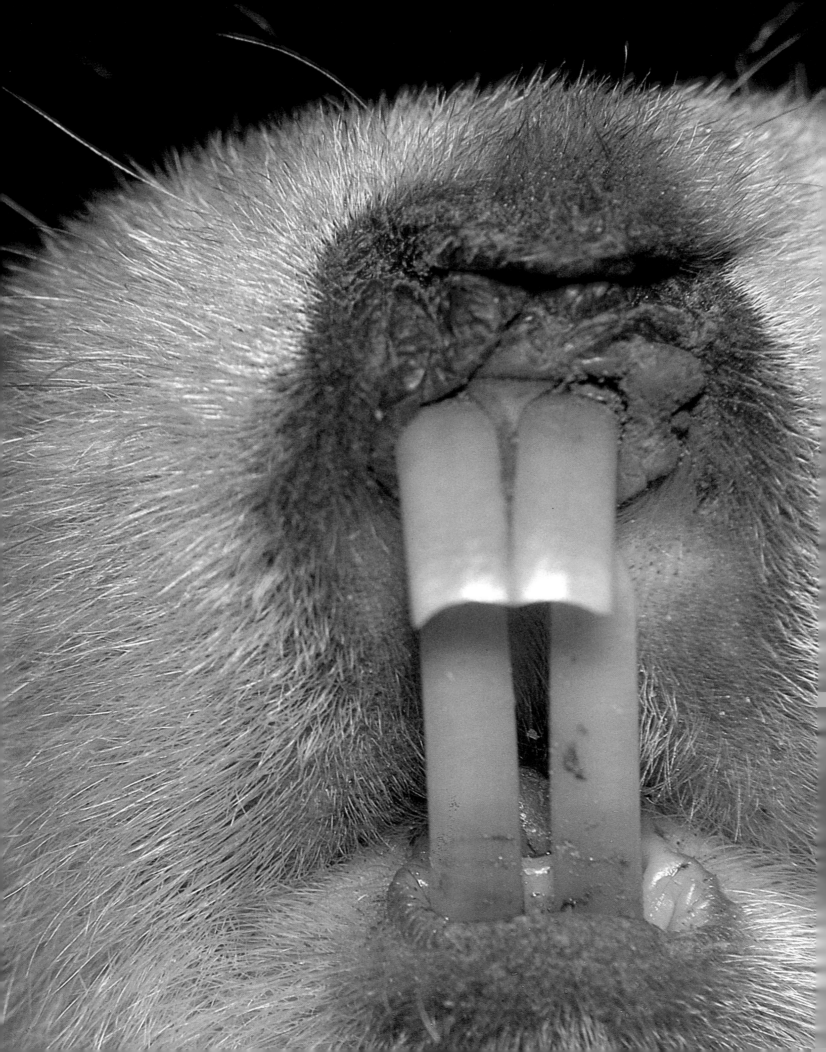

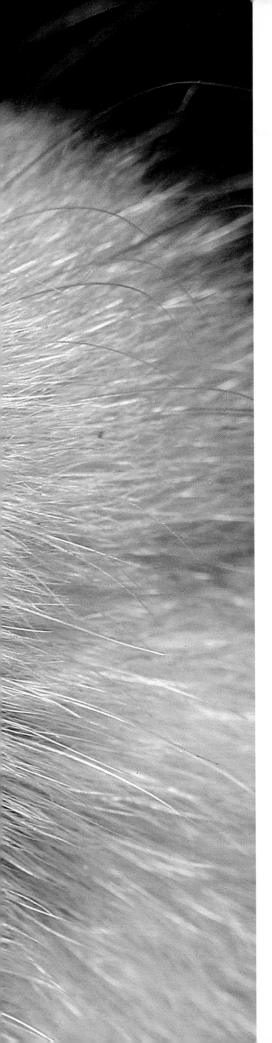

FOREWORD

Rodents is a fascinating and illuminating book, full of surprises for those who until now have only associated these small herbivores with the Plague, crop devastation and mouse droppings in the kitchen. Many of the world's 1,775 rodent species are undeniably destructive and pose enormous threats to man, but their poor public image is unworthy of an order of mammals that is as diverse as it is interesting. Besides rats and mice, rodents include squirrels, beavers, gophers, hamsters, porcupines and cavies.

Not only do rodents of one species or another occupy almost every nook and cranny available on dry land, they do so with conspicuous success, despite strenuous human efforts to eliminate them. For not only are they able to reproduce with astonishing haste and profligacy, rodents are also impressively and highly adaptable. Many rodents do not conform at all to human preconceptions. Some burrow like moles, others fish in streams and yet others climb tall trees in search of food. Some South American rodents are the size of sheep. *Rodents* gives a graphic insight into their societies which, as these pages illustrate, are no less varied than their bodies.

A second major group of small herbivores—the lagomorphs—is also discussed and illustrated in these pages. They share most of the characteristics of rodents—global distribution, small size and incredible breeding capacity, not to mention, in some cases, a similar destructive potential. "Lagomorph" means "hare-shaped" and while most members of this group do indeed conform to the "bunny" stereotype, others—the pikas from North America and Asia—actually look more like hamsters.

The book concludes with a study of elephant-shrews. These creatures, which are only found in Africa, have snouts like elephants and the body form of a small antelope with a ratlike tail. Their lifestyles are equally as intriguing and unusual.

How this book is organized

Animal classification can be a thorny problem—one on which even the experts sometimes find it difficult to agree. This book has taken the views of many taxonomists, but in general it follows the classification of Corbet and Hill (see Bibliography).

Rodents is structured on three main levels. First there is a general essay which highlights the common features and main variations of the biology (particularly the body plan), ecology, behavior, and evolution of rodents. Second, essays on taxonomic groupings—generally orders, suborders and families—of rodents, lagomorphs and elephant-shrews highlight topics of particular interest. Each of these essays is introduced with a summary of species or species groupings, a description of the skull, dentition and any unusual skeletal features of representative species, plus a map showing distribution. Third, the main text, devoted to individual species or groups of species, covers details of physical features, distribution, evolutionary history, diet and feeding behavior. The animals' social dynamics, spatial organization, conservation and relationship with man are also discussed.

Before the textual discussion of each species or group, readers will find an information panel. This gives basic data about size, lifespan and distribution. A map shows areas of natural distribution, while a scaled drawing compares the size of each species with a six-foot man or, where more appropriate, with part of a six-foot man or a twelve-inch human foot. When silhouettes of two animals are shown, they are the largest and the smallest representatives of the group. If the panel covers a large group of species, those listed as examples in the panel are those referred to in the text. Detailed descriptions of the remaining members of a group are to be found in a separate Table of Species. Unless otherwise stated, dimensions given are for both males and females. Where there is a difference in size between the sexes, the scale drawings show males.

In many instances the text is enhanced and amplified by specially commissioned color artwork showing representative species engaged in characteristic activities. Simpler line drawings illustrate particular aspects of behavior or clarify anatomical distinctions between otherwise similar species. Color photographs show creatures in their natural habitats and, where possible, displaying typical behavior.

While many larger herbivores are listed and protected as endangered species, measures for conservation are clearly unnecessary in the case of rodents. Of the total of 1,775 species that comprise the families considered in these pages, only four are listed in either CITES (Appendices I through III of the Convention on International Trade in Endangered Species of Wild Flora and Fauna) or in the *Red Data Book* of the International Union for the Conservation of Nature and Natural Resources. All four are from the lagomorph family and include the Sumatran hare, of which only twenty individuals have ever been recorded. The symbol Ⓔ has been used to show that the status accorded to a species by the IUCN at the time of going to press is Endangered. This means that the species is in danger of extinction unless causal factors are modified. The symbol ⊡ indicates species listed by CITES.

RODENTS

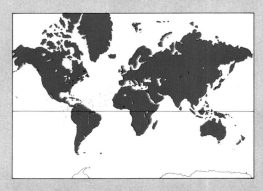

ORDER: RODENTIA
Thirty families: 389 genera: 1,702 species.

Squirrel-like rodents
Suborder: Sciuromorpha—sciuromorphs
Seven families: 65 genera: 377 species.

Beavers
Family: Castoridae
Two species in 1 genus.

Mountain beavers
Family: Aplodontidae
One species.

Squirrels
Family: Sciuridae
Two hundred and sixty-seven species in 49 genera.
Includes **Bryant's fox squirrel** (*Sciurus niger*), **flying squirrels** (subfamily *Petaurista*), **Gray squirrel** (*Sciurus carolinensis*), **ground squirrels** (genus Spermophilus), **marmots** (genus *Marmota*), **prairie dogs** (genus *Cynomys*), **Red squirrel** (*Sciurus vulgaris*), **Woodchuck** (*Marmota monax*).

Pocket gophers
Family: Geomyidae
Thirty-four species in 5 genera.

Scaly-tailed squirrels
Family: Anomaluridae
Seven species in 3 genera.

Pocket mice
Family: Heteromyidae
Sixty-five species in 5 genera.
Includes **Big-eared kangaroo rat** (*Dipodomys elephantinus*), **Kangaroo mice** (genus *Microdipodops*).

Springhare or springhass
Family: Pedetidae
One species.

Mouse-like rodents
Suborder: Myomorpha—myomorphs
Five families: 264 genera: 1,137 species.

Rats and mice
Family: Muridae
One thousand and eighty two species in 241 genera and 15 subfamilies.
Includes: **Australian water rat** (*Hydromys chrysogaster*), **Brown lemmings** (genus *Lemmus*), **Collared lemming** or **Arctic lemming** (*Lemmus torquatus*), **Common vole** (*Microtus arvalis*), **cotton rats** (genus *Sigmodon*), **Golden hamster** (*Mesocricetus auratus*), **House mouse** (*Mus musculus*), **Multimammate rat** (*Praomys (Mastomys) natalensis*), **Nile rat** (*Arvicanthis niloticus*), **Norway lemming** (*Lemmus lemmus*), **Norway** or **Common** or **Brown rat** (*Rattus norvegicus*), **Roof rat** (*Rattus rattus*), **Saltmarsh**

Rodents have influenced history and human endeavor more than any other group of mammals. Nearly 40 percent of all mammal species belong to this one order, whose members live in almost every habitat, often in close association with man. Frequently this association is not in man's interest, for rodents consume prodigious quantities of his carefully stored food, and spread fatal diseases. It is said that rat-borne typhus has been a greater influence upon human destiny than any single person, and that in the last millennium rat-borne diseases have taken more lives than all wars and revolutions put together. Rodents have never been highly beneficial to man, though some larger species have been, and still are, sought for food in many parts of the world. Only a handful of species, such as guinea pigs and edible dormice, have been deliberately bred for food. Ironically, rats, mice and guinea pigs today play an inestimable role as "guinea pigs" in the testing of drugs and in biological research.

The reason for this ambivalent relationship is that rodents are as highly adaptable as man and, like man, are opportunists. Our shared adaptability has guaranteed conflict of interest. In expanding our own world we invariably opened new habitats for our rodent competitors.

The success of the rodents is partly attributable to the fact that in evolutionary terms, they are quite young (between 26 and 38 million years), and populations retain large, untapped stocks of genetic variability. This variability is exposed to the selective forces of evolution rapidly since rodents produce many large litters each year. They are able to try out quickly new genetic combinations in the face of new environmental conditions (see pp80–81). A second facet of their success is that rodents have a very wide ranging diet.

Rodents occur in every habitat, from the high arctic tundra, where they live and breed under the snow (eg lemmings), to the hottest and driest of deserts (eg gerbils). Others glide from tree to tree (eg flying squirrels), seldom coming down to the ground, while others spend their entire lives in an underground network of burrows (eg mole-rats). Some have webbed feet and are semiaquatic, (eg muskrats), often undertaking complex engineering programs to regulate water levels (eg beavers), while others never touch a drop of water throughout their entire lives (eg gundis). Such species can derive their water requirements from fat reserves.

Rodents show less overall variation in body plan than do members of many mammalian orders. Most rodents are small, weighing 3.5oz (100g) or less. There are only a few large species of which the largest, the capybara, may weigh up to 146lb (66kg). All rodents have characteristic teeth, including a single pair of razor-sharp incisors. With these teeth the rodent can gnaw through the toughest of husks, pods

▶ **The expected build of a rodent.** This bush rat (*Rattus fuscipes*) would be recognizable as a rat to anyone familiar with the world's main rats, the Norway (or Common) and Roof rats, which are found wherever humans live. This rat lives along the eastern and southern coasts of Australia.

▼ **A highly adapted rat.** Many rodents are adapted for swimming, including this Australian water rat which is found in all nonarid areas of Australia. Its fur is seal-like and waterproof, its face is streamlined, its ears and eyes are small, its hindfeet webbed. It lives in marshes, swamps, backwaters and rivers, as do many of the world's rodents. It is also primarily carnivorous.

harvest mouse (*Reithrodontomys raviventris*),
Wood mouse (*Apodemus sylvaticus*).

Dormice
Families: Gliridae, Seleviniidae
Eleven species in 8 genera.

Jumping mice and birchmice
Family: Zapodidae
Fourteen species in 4 genera.

Jerboas
Family: Dipodidae
Thirty-one species in 11 genera.

Cavy-like rodents
Suborder: Caviomorpha—caviomorphs
Eighteen families: 60 genera: 188 species.

New World porcupines
Family: Erethizontidae
Ten species in 4 genera.

Cavies
Family: Caviidae
Fourteen species in 5 genera.

Capybara
Family: Hydrochoeridae
One species.

Coypu
Family: Myocastoridae
One species.

Hutias
Family: Capromyidae
Thirteen species in 4 genera.

Pacarana
Family: Dinomyidae
One species.

Pacas
Family: Agoutidae
Two species in 1 genus.

Agoutis and acouchis
Family: Dasyproctidae
Thirteen species in 2 genera.

Chinchilla rats
Family: Abrocomidae
Two species in 1 genus.

Spiny rats
Family: Echimyidae
Fifty-five species in 15 genera.

Chinchillas and viscachas
Family: Chinchillidae
Six species in 3 genera.

Degus or Octodonts
Family: Octodontidae
Eight species in 5 genera.

Tuco-tucos
Family: Ctenomyidae
Thirty-three species in 1 genus.

Cane rats
Family: Thryonomyidae
Two species in 1 genus.

African rock rat
Family: Petromyidae
One species.

Old World porcupines
Family: Hystricidae
Eleven species in 4 genera.

Gundis
Family: Ctenodactylidae
Five species in 4 genera.

African mole-rats
Family: Bathyergidae
Nine species in 5 genera.

and shells to secure the nutritious food contained within. The name "rodent" comes from the Latin *rodere*, which means to gnaw. Gnawing is facilitated by a sizable gap, called the diastema, immediately behind the incisors, into which the lips can be drawn, so sealing off the mouth from inedible fragments dislodged by the incisors. Rodents have no canine teeth, but they do possess a substantial battery of molar teeth by which all food is finely ground. These often massive and complexly structured teeth are traversed by convoluted layers of enamel. The pattern made by these layers is often of taxonomic significance. Most rodents have no more than 22 teeth, though one exception is the Silvery mole-rat from Central and East Africa which has 28. The Australian water rat has just 12. Since rodents feed on hard materials, the incisors have open roots and grow continuously throughout life; as much as an eighth of an inch per week. They are constantly worn down by the action of their opposite number on the other jaw. If the teeth of rodents become misaligned so that they are not automatically worn down they will grow around and pierce the skull.

Most rodents are squat, compact creatures with short limbs and a tail. In South America, where there are no antelopes, several species have evolved long legs for a life on the grassy plains (eg maras, pacas and agoutis), and show some convergence towards the antelope body form. A very variable anatomical feature is the tail (see opposite).

The order is divided into three on the basis of the arrangement of jaw muscles. The principal jaw muscle is the masseter muscle which not only closes the lower jaw on the upper, but also pulls the lower jaw forward so allowing the gnawing action to occur. This action is unique. In the extinct Paleocene rodents the masseter was small and did not spread far onto the front of the skull. In the squirrel-like rodents (Sciuromorpha) the lateral masseter extends in front of the eye onto the snout; the deep masseter is short and used only in closing the jaw (see p20). In the cavy-like rodents (Caviomorpha) it is the deep masseter that extends forwards onto the snout to provide the gnawing action (see p100). Both the lateral and deep branches of the masseter are thrust forward in the mouse-like rodents (Myomorpha), providing the most effective gnawing action of all, with the result that they are the most successful in terms of distribution and number of species (see pp52–99).

Most rodents eat a range of plant products, from leaves to fruits, and will also eat small invertebrates, such as spiders and grasshoppers. A few are specialized carnivores; the Australian water rat feeds on small fish, frogs and mollusks and seldom eats plant material. To facilitate bacterial digestion of cellulose rodents have a relatively large cecum (appendix) which houses a dense bacterial flora (though this structure is relatively smaller in dormice). After the food has been softened in the stomach it passes down the large intestine and into the cecum. There the cellulose is split by bacteria into its digestible carbohydrate constituents, but absorption can only take place higher up the gut, in the stomach. Therefore rodents practise refection—reingesting the bacterially treated food taken directly from the anus. On its second visit to the stomach the carbohydrates are absorbed and the fecal pellet that eventually emerges is hard and dry. It is not known how rodents know which type of feces is being produced. The rodent's digestive system is very efficient, assimilating 80 percent of the ingested energy.

All members of at least three families (hamsters, pocket gophers, pocket mice) have cheek pouches. These are folds of skin, projecting inwards from the corner of the mouth, and are lined with fur. They may reach back to the shoulders. They can be everted for cleaning. Rodents with cheek pouches build up large stores—up to 198lb (90kg) in Common hamsters.

Rodents are intelligent and can master simple tasks for obtaining food. They can be readily conditioned, and easily learn to avoid fast-acting poisoned baits—a factor that makes them difficult pests to eradicate (see pp18–19). Their sense of smell and their hearing are keenly developed. Nocturnal species have large eyes and all rodents have long, touch-sensitive whiskers (vibrissae).

In fact two-thirds of all rodent species belong to just one family, the Muridae, with 1,082 species. Its members are distributed worldwide, including Australia and New Guinea, where it is the only terrestrial placental mammal family to be found (excluding modern introductions such as the rabbit). The second most numerous family is that of the squirrels (Sciuridae) with 267 species distributed throughout Eurasia, Africa and North and South America.

The fossil record of the rodents is pitifully sparse. Rodent remains are known from as far back as the late Paleocene era (57 million years ago), when all the main characteristics of the order had developed. The earliest

THE RODENT BODY PLAN

▼ **Skull of the Roof rat.** Clearly shown are the continuously growing, gnawing incisors and the chewing molars, with the gap (diastema) left by the absence of the canine and premolar teeth. All mouse-like rodents lack premolars but the squirrel- and cavy-like rodents have one or two on each side of the jaw.

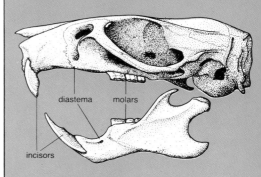

diastema   molars

incisors

▶ **Skeleton of the Roof rat.** This is typical of rodents with its squat form, short limbs, plantigrade gait (ie it walks on the soles of its feet) and long tail.

▶ **Evolutionary relationships** within the order Rodentia. The chart shows both the extent of the known fossil record (plum) and hypothetical lines of descent (pink).

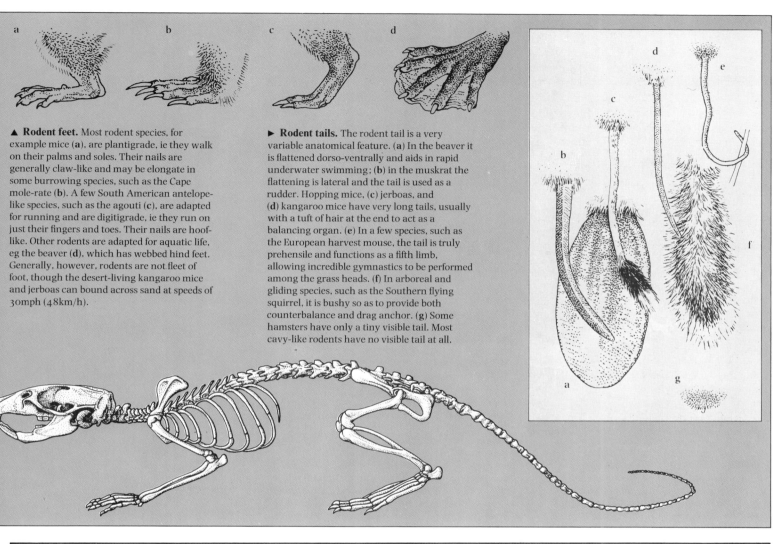

▲ **Rodent feet.** Most rodent species, for example mice (**a**), are plantigrade, ie they walk on their palms and soles. Their nails are generally claw-like and may be elongate in some burrowing species, such as the Cape mole-rate (**b**). A few South American antelope-like species, such as the agouti (**c**), are adapted for running and are digitigrade, ie they run on just their fingers and toes. Their nails are hoof-like. Other rodents are adapted for aquatic life, eg the beaver (**d**), which has webbed hind feet. Generally, however, rodents are not fleet of foot, though the desert-living kangaroo mice and jerboas can bound across sand at speeds of 30mph (48km/h).

▶ **Rodent tails.** The rodent tail is a very variable anatomical feature. (**a**) In the beaver it is flattened dorso-ventrally and aids in rapid underwater swimming; (**b**) in the muskrat the flattening is lateral and the tail is used as a rudder. Hopping mice, (**c**) jerboas, and (**d**) kangaroo mice have very long tails, usually with a tuft of hair at the end to act as a balancing organ. (**e**) In a few species, such as the European harvest mouse, the tail is truly prehensile and functions as a fifth limb, allowing incredible gymnastics to be performed among the grass heads. (**f**) In arboreal and gliding species, such as the Southern flying squirrel, it is bushy so as to provide both counterbalance and drag anchor. (**g**) Some hamsters have only a tiny visible tail. Most cavy-like rodents have no visible tail at all.

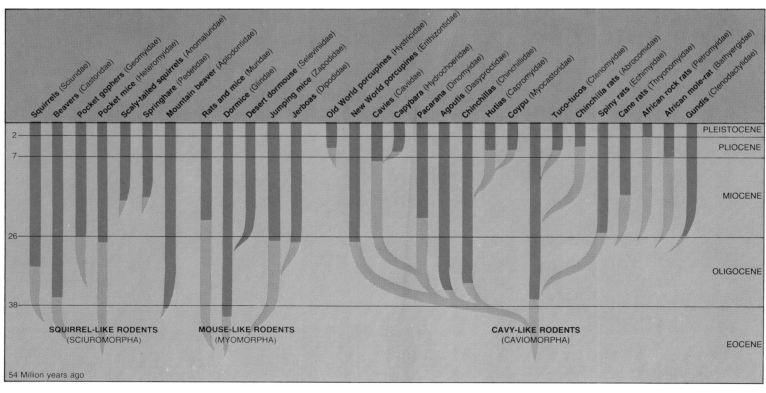

SQUIRREL-LIKE RODENTS
(SCIUROMORPHA)

MOUSE-LIKE RODENTS
(MYOMORPHA)

CAVY-LIKE RODENTS
(CAVIOMORPHA)

Squirrels (Sciuridae)
Beavers (Castoridae)
Pocket gophers (Geomyidae)
Pocket mice (Heteromyidae)
Scaly-tailed squirrels (Anomaluridae)
Springhare (Pedetidae)
Mountain beaver (Aplodontidae)
Rats and mice (Muridae)
Dormice (Gliridae)
Desert dormouse (Seleviniidae)
Jumping mice (Zapodidae)
Jerboas (Dipodidae)
Old World porcupines (Hystricidae)
New World porcupines (Erithizontidae)
Cavies (Caviidae)
Capybara (Hydrochoeridae)
Pacarana (Dinomyidae)
Agoutis (Dasyproctidae)
Chinchillas (Chinchillidae)
Hutias (Capromyidae)
Coypu (Myocastoridae)
Tuco-tucos (Ctenomyidae)
Chinchilla rats (Abrocomidae)
Spiny rats (Echimyidae)
Cane rats (Thryonomyidae)
African rock rats (Petromyidae)
African mole-rat (Bathyergidae)
Gundis (Ctenodactylidae)

PLEISTOCENE
2
PLIOCENE
7
MIOCENE
26
OLIGOCENE
38
EOCENE

54 Million years ago

rodents belonged to the extinct sciuromorph family Paramyidae. During the Eocene era (54–38 million years ago) there was a rapid diversification of the rodents, and by the end of that epoch it seems that leaping, burrowing and running forms had evolved. At the Eocene–Oligocene boundary (38 million years ago) many families recognizable today occurred in North America, Europe and Asia, and during the Miocene (about 26 million years ago) the majority of present-day families had arisen. Subsequently the most important evolutionary event was the appearance of the Muridae in the Pliocene era (7 million years ago) from Europe. About 2.5 million years ago, at the start of the Pleistocene era, they entered Australia, probably via Timor, and then underwent a rapid evolution. At the same time, when North and South America were united by a land bridge, murids invaded from the north with the result that there was an explosive radiation of New World rats and mice in South America (see p56).

The Role of Odor in Rodent Reproduction

Reproduction—from initial sexual attraction and the advertisement of sexual status through courtship, mating, the maintenance of pregnancy and the successful rearing of young—is influenced, if not actually controlled, by odor signals.

Male rats are attracted to the urine of females that are in the sexually receptive phase of the estrous cycle and sexually experienced males are more strongly attracted than naive males. Furthermore, if an experienced male is presented with the odor of a novel mature female alongside the odor of his mate he prefers the novel odor. Females, on the other hand, prefer the odor of their stud male to that of a stranger. The male's reproductive fitness is best suited by his seeking out, and impregnating, as many females as possible. The female needs to produce many healthy young, so her fitness is maximized by mating with the best quality stud male who has already proved himself. The otherwise solitary female Golden hamster must attract a male when she is sexually receptive. This she does by scent marking with strong-smelling vaginal secretions in the two days before her peak of receptivity. If no male arrives she ceases marking, to start again two days before the next peak.

In gregarious species such as the House mouse a dominant male can mate with 20 females in 6 hours if their cycles are synchronized. The odor of urine of adult sexually mature male rodents (eg mice, voles, deermice) accelerates not only the peak of female sexual receptivity but also the onset of sexual maturity in young females and brings sexually quiescent females into breeding condition. The effect is particularly strong from the urine of dominant males. Urine from castrated males has no such effect, so it would appear that the active ingredient—a pheromone—is made from, or dependent upon the presence of, the male sex hormone testosterone. Male urine has such a powerful effect that if a newly pregnant female mouse is exposed to the urine odor of a male who is a complete stranger to her she will resorb her litter and come rapidly into heat. If she then mates with the stranger she will become pregnant and carry the litter to term. The odor of the urine of females has either no effect upon the timing of the onset of sexual maturity in young females, or slightly retards it. If female mice are housed together in groups of 30 or more and males are absent the normal 4- or 5-day estrous cycles start to lengthen and the incidence of pseudopregnancy increases, indicating the power of the odor of female urine. However, the presence of the urine odor of an adult male will regularize all the lengthened cycles within 6–8 hours and the females will come into heat synchronously.

Female mice also produce a pheromone in the urine which has the effect of stimulating pheromone production in the male, but the female pheromone is not under the control of the sex glands (ovaries). It is not known what does control its production. A sexually quiescent female could stimulate pheromone production in a male, which would then bring her into sexual readiness.

It is thought that the reproductive success of the House mouse owes much to this system of pheromonal cuing, for mice would never be faced with a situation in which neither sex could stimulate or respond to the other because both stimulus and response were dependent upon the presence of fully functional sex glands in both sexes. Although only the House mouse has been studied in such detail, parts of the model have been discovered in other species and it may be of widespread occurrence.

About 8 days after giving birth, female rats start to produce a pheromone—an odor produced in the gut and broadcast via the feces—which inhibits *Wanderlust* in the young. It ceases to be produced when the young are 27 days old and almost weaned.

Finally some studies have involved a surgical removal of part of the brain which is involved with smell (the main and accessory olfactory bulbs). Removal of the bulbs in the Golden hamster, irrespective of previous sexual experience, bring an immediate cessation of all sexual behavior. In sexually experienced rats the operation has little effect, but in sexually naive rats the effect is as severe as in hamsters. Thus it appears that rats can learn to do without their sense of smell once they have gained some sexual experience.

DMS

▲ **Portable shade in the desert.** These African ground squirrels (*Xerus inauris*), which live in the Kalahari desert, provide themselves with shade by fluffing out their tails.

◄ **Threat displays** are very dramatic in some rodent species. (1) When slightly angry the Cape porcupine raises its quills and rattles the specialized hollow quills on its tail. If this fails to have the desired effect the hind feet are thumped on the ground in a war dance accompaniment to the rattling. Only if the threat persists will the porcupine turn its back on its opponent and charge backwards with its lethal spines at the ready. (2) Slightly less dramatic is the threat display of the Kenyan crested rat. This slow and solidly built rodent responds to danger by elevating a contrastingly colored crest of long hairs along its back, and in so doing exposes a glandular strip along the body. Special wick-like hairs lining the gland facilitate the rapid dissemination of a strong, unpleasant odor. (3) Finally, the little Norway lemming stands its ground in the face of danger and lifts its chin to expose the pale neck and cheek fur which contrast strongly with the dark upper fur.

With their high powers of reproduction and ability to invade all habitats, rodents are of great ecological importance. Occasionally their numbers become astoundingly high—almost 80,000 House mice per acre (200,000 per ha) were recorded in Central Valley in California in 1941–42—with densities of up to 500 per acre (1,200 per ha) for grassland species such as the Common vole being common. Apart from the immediate economic ruin that such hordes may bring to farmers, such densities have a profound effect upon the ecological balance of an entire region. Firstly, considerable damage is inflicted on vegetation resulting from the wholesale deaths of basic plants, from which it may take several years to recover. Secondly, the abundance of predators increases in response to the abundance of rodents, and when the rodents have gone they turn their attention to other prey. Such dramatic densities seem to occur only in unpredictable environments (but not all), such as the arctic tundra, and in marginal farming regions of arid zones. The numbers of tropical rodents, for the most part, rarely in-

crease such that they become a nuisance to man.

A characteristic type of population explosion is seen in many rodents of the arctic tundra and taiga. In these areas population explosions occur with regularity, every 3–4 years, involving the Norway lemming in Europe and the Brown and Collared lemmings in North America. The population density builds up to a high level and then dramatically declines. A number of ideas have been put forward to explain the decline, but they are all too simplistic. For example, it was long held that the decline was brought about by disease (tularaemia or lemming fever which is found in many rodent populations), but it seems more likely that disease simply hastens the decline. Another suggestion was that the rodents become more aggressive at high density, which leads to a failure of courtship and reproduction. Others include the action of predators, and the impoverishment of the forage. Objective observation of lemming behavior at high density shows it to be adaptive, providing the lemmings with the best chances for survival (see p70).

Rodent cycles would not occur if rodents were not such prodigious breeders. Norway lemmings, for example, mate when they are about 4 weeks old and just newly weaned. Litters of about 6 young can be produced at 3–4-week intervals. Ecologists regard rodents as "r" selected, ie they are adapted for rapid reproductive turnover. Throughout mammals, species with short lives reproduce rapidly while long-lived species reproduce more slowly.

Although some species of rodents are extremely rare (eg the Big-eared kangaroo rat from Central California, *Dipodomys elephantinus*), few are listed in the *Red Data Book* as being endangered, only the Vancouver Island marmot, the Delmarva fox squirrel and the Saltmarsh harvest mouse (from marshes of southern California). A number of species are under threat, including six species of hutias (Family Capromyidae) from the islands of the West Indies, and in all these cases the danger comes from the draining of land and the clearing of formerly dense bush.

Rodents are highly social mammals, frequently living in huge aggregations. Prairie dog townships may contain more than 5,000 individuals. The solitary way of life appears to be restricted to those species that live in arid grasslands and deserts—hamsters and some desert mice—and to a few strictly territorial species such as the woodchuck. Their behavior lacks the ability to switch off aggression in others—they have no capacity for appeasement. Observation of a cage of mice for just a few moments will indicate the frequency of appeasement signals: a turning away of the head, a closing of the eyes. Fighting never occurs when one of a pair shows appeasement.

Rodent behavior is as adaptable as every other aspect of rodent biology. For example, individual House mice may be strongly territorial when population density is low, with adult males dribbling urine around their territories to saturate it with the owner's odor, but when density is high the territorial system breaks down in favor of a social hierarchy, or "pecking order." The consequences of both systems are the same: discrimination between the "haves" and the "have-nots" with regard to allocated limited resources. Dominant individuals display a confident deportment and show little appeasement. They are frequently the oldest and largest male members of the population. The blood level of the male sex hormone testosterone is higher in higher ranking individuals. Some species have extremely complex systems of social organization, for example Naked mole rats (see pp126–27), but for the great bulk of mouse- and rat-sized species little is known.

Having alert and active senses, rodents communicate by sight, sound and smell. Some of the best known visual displays are seen in the arboreal and the day-active terrestrial species. Courtship display in tree squirrels may be readily observed in city parks in early spring. The male pursues the female through the trees, flicking his bushy tail forward over his body and head when he is stationary. The female goads him by running slightly ahead, but he responds by uttering infantile "contact-keeping" sounds similar to those infants produce to keep their mothers close. These stop the female, allowing the male to catch up. Threat displays are dramatic in some species (see pp14–15).

Rodents make considerable use of vocalizations in their social communication. Occasionally, as in the threat of a lemming, the sound is audible (ie below 20kHz), but more often frequencies are far above the range of human hearing (at about 45kHz).

Rodents communicate extensively through odors which are produced by a variety of scent glands (see p14). Males tend to produce more and stronger odors than females, and young males are afforded a measure of protection from paternal attack by smelling like their mothers until they are sexually mature.   DMS

Rodents as pests

Of the approximately 1,700 species of rodents only a handful are economically important, of which some occur worldwide. Recently the Food and Agriculture Organization (FAO) of the United Nations estimated that each year rodent pests worldwide consume 42.5 million tons of food, worth 30 billion US dollars, equivalent to the gross national product of the world's 25 poorest countries. In addition, rodent pests are involved in the transmission of more than 20 pathogens, including plague (actually transmitted to man by the bite of the rat flea). Bubonic plague was responsible for the death of 25 million Europeans from the 14th century to the 17th.

The principal rodent pests are three species of the rats and mice family: the Norway, Common or Brown rat, the Roof rat and the House mouse. Squirrel-like pests include the European red and Gray squirrels which are, in terms of the cost of their damage, relatively minor pests (see pp42–43). In 1974, for example, only 1 percent of the 48 percent of trees susceptible in the United Kingdom were damaged. Other pests include the Common hamster,

▲ **Dynamic reproductive power** is latent in the major rodent pests. Female Norway rats, for example, can breed when only two months old and can produce up to 11 young in one litter, and give birth again three or four weeks later.

▶ **A worldwide problem.** Vast populations of the major rodent pests occur worldwide, and can mobilize themselves with devastating results. From top to bottom: rat damage to orange in Egypt; a citrus tree stripped by rats in Cyprus; sugar cane damaged by the Hispid cotton rat. In the bottom picture of a sugar cane plantation all but the dark green areas was destroyed by rats.

which is actually common in central Europe and is a pest on pasture land, and the gerbils which devastate agricultural crops in Africa. In North America, eastern and western Europe voles are prominent pests: they strip bark from trees, often causing death, and consume seedlings in forest plantations or fields. When vole populations peak (once every 3 or 4 years) there may be 5,000 voles per acre (almost 2,000 per ha). Other major rodent pests include the Cotton or Cane rat and *Holochilus braziliensis*, from, principally, Latin America, the Nile rat and the Multimammate rat of North and East Africa, *Rattus exulans* of the Pacific Islands, the *Bandicota* rats of the Indian subcontinent and Malaysian Peninsula and the prairie dogs, marmots and ground squirrels of the Mongolian and Californian grasslands.

The characteristics that enable these animals to become such major pests are a simple body plan, adapted to a variety of habitats and climates; high reproductive potential; opportunistic feeding behavior; and gnawing and burrowing habits.

In Britain the Norway rat may live in fields during the warm summer months where food is plentiful and they seldom reach economically important numbers. However, after harvest and with the onset of cold weather they move into buildings. Also in Britain, and some other western European countries, the wood mouse *Apodemus sylvaticus*, normally only a pest in the winter when it may nibble stored apples for example, has learnt to locate, probably by smell, pelleted sugar beet seed (now commonly precision drilled to avoid the need for hoeing out the extra production of seedling plants which used to be allowed for). This damage, possibly occurring unnoticed

before the advent of precision drilling, can lead to large barren patches in fields of sugar beet and sometimes necessitates complete resowing.

Some rodents cause damage to agricultural crops, stored products and structural components of buildings etc. Such hazards result from rodents' feeding behavior or indirectly by their gnawing and burrowing. Although rodents usually consume about 10 percent of their body weight in food per day, much of the damage they do is not due to direct consumption. Three hundred rats in a grain store can eat 3 tons of grain in a year, but every 24 hours they also produce and contaminate the grain with 15,000 droppings, 6 pints (3.5 litres) of urine and countless hairs and greasy skin secretions. In sugar cane, rats may chew at the cane directly consuming only a part of it. The damage, however, may cause the cane to fall over (lodge) and the impact of the sun's rays will be reduced and harvesting impeded. The gnawed stem allows microorganisms to enter reducing the sugar content. Apart from the value of the lost crop, if the result is a loss in the sugar content of the crop of 6 percent that represents 6 percent of the investments in land preparation, fertilizers, pesticides, irrigation water, management, harvesting and processing. In Asia rodents routinely consume 5–30 percent of the rice crop and not infrequently devastate whole areas of rice fields, sometimes in excess of 25,000 acres (10,000ha). Rice field rats eat seedling rice and cut the shoots of rice as they ripen to consume the dough or milky stage of developing rice seedheads. A small amount of early damage may be compensated for by extra growth of the plant, but even if this happens the extra growth may ripen too late for harvesting. In oil palm in West Africa and especially Malaysia rats eat the hearts out of very young oil palm trees and consume the flesh of the oil palm fruit leading to losses in Malaysia alone of 5 percent of the final oil palm yield, valued at about 50 million US dollars per year.

Tropical crops damaged by rodents include coconuts, maize, coffee, field beans, citrus, melons, cocoa and dates. Each year rodents consume food equivalent to the total of all the world's cereal and potato harvest; a train 3,000mi long, (about 5,000km) as long as the Great Wall of China, would be needed to haul all this food.

Rodents also cause considerable losses of stored grain. For example, China produces about 320 million tons of grain a year and has to import a further 11 million tons to

feed its 1,008 million population. Every year rodents, particularly the House mouse and Norway rat, are responsible for losses from this stored grain, amounting to at least 5–10 percent of the total, most from farm or village stores. The prevention of the loss, which could be achieved in a cost-effective manner, would eliminate the need for grain imports.

Structural damage attributable to rodents results, for example, from the animals burrowing into banks, sewers and under roads. The effects include subsidence, flooding or even soil erosion in many areas of the world. Gnawed electrical cables can cause fires, with their enormous economic impact. In the insulated walls of modern poultry units rodents will gnaw through electrical wires causing malfunctions of air-conditioning units, and subsequent severe economic losses. In Sudan, for example, a single power loss of 1 hour resulted in overheating of one unit causing the death of 8,000 birds to the value of 20,000 US dollars.

It is extremely difficult to put a value on the effects of diseases transmitted by rodents. There is little accurate information, partly due to a dearth of specialists in the field and partly because many governments do not wish to reveal the problems of disease in their countries in case it adversely affects their tourist industry.

Apart from plague, which persists in many African and Asian countries as well as in the USA (where wild mammals transmit the disease killing fewer than 10 people a year), murine typhus, food poisoning with *Salmonella*, Weils disease (Leptospirosis) and the West African disease Lassa fever, to mention just a few, are potentially fatal diseases transmitted by rats. Rats are probably responsible for more human deaths than all wars and revolutions. Difficult though it is to put an economic value on loss of life it is even more difficult to evaluate the economic consequences of debilitating chronic disease reducing the working efficiency of whole populations. One developing Asian country's capital was recently found to have 80 percent of its population seropositive for murine typhus, ie they had all, sometime, been in contact with the disease and had suffered from at least a mild form of it. Forty percent of people admitted to hospital in the capital city were diagnosed as having fever of unknown origin, at least some of these, maybe a majority, were probably suffering from murine typhus. The impact of this disease on the economy of the country is impossible to determine and the same country also suffered frequent out-

breaks of plague. Rats are involved in the transmission of both diseases.

To control the impact of rodents the simplest method is to reduce harborage and available food and water. This, however, is often impossible. At best such "good housekeeping" can prevent rodent populations building up, but it is seldom an effective method for reducing existing rodent populations. The same principles of management can be applied to fields, stores or domestic premises. In buildings, food and organic rubbish must be made inaccessible to rodents by blocking nooks and crannies where rodents may find refuge. They must be denied access to underground ducting (including pipe runs and sewers) as well as to air-conditioning ducts.

To reduce existing populations of rodents, predators—wild or domestic (such as cats or dogs)—have relatively little effect. Their role may lie in limiting population growth. It is a widely accepted principle that predators do not control, in absolute terms, their prey although the abundance of prey may effect the numbers of predators. Mongooses were introduced to the West Indies and Hawaiian Islands and Cobra snakes to Malaysian oil palm estates, both for controlling rats. The rats remain and the mongooses and cobra are now considered pests. Even the farm cat will not usually have a significant effect on rodent numbers: the reproductive rate of rodents keeps them ahead of the consumption rate of cats!

One of the simplest methods for combating small numbers of rodents is the use of traps. Few, however, are efficient: most just maim their victims. The most efficient method of control is by modern chemical rodenticides. The oldest kind of rodenticide, fast-acting, nonselective poison, appears in the earliest written record of chemical pest control. Aristotle, in 350 BC, described the use of the poison strychnine. However, fast-acting poisons (such as cyanide, strychnine, sodium monofluoroacetate (compound 1080), thallium sulfate and zinc phosphide) have various technical and ecological disadvantages, including long-lasting poison shyness in sublethally poisoned rodents and high hazards to other animals, and are not normally recommended today.

Since 1945, when warfarin was first synthesized, several anticoagulant rodenticides have been developed. These compounds decrease the blood's ability to clot, and lead to death by internal or external bleeding. These compounds have some selectivity as most have to be ingested by rodents over a number of days (5–10) for a

▲ **Rodents' impact on human history.** Numerous major outbreaks of rodent-borne disease in history have caused economic dislocation and social change on an enormous scale. Here victims of the Great Plague in London (1664–5) are buried. They died of *Pasteurella pestis*, transmitted by fleas carried by rats.

▶ **Rat in a grain store:** a Norway rat.

▼ **Communal destruction:** a night time scene in an Australian grain store.

lethal dose to be achieved; vitamin K_1 is used as an antidote in the case of accidental poisoning. Such compounds are best suited to controlling the Norway rat but used properly can result in total control of infestations of susceptible species. In some countries rodents, particularly the House mouse and Norway rat, developed genetic resistance to these "first generation" anticoagulants, notably in the United Kingdom, USA, Canada, Denmark and France, where the products have been used intensively for 10 years or more. This led to the development of "second generation" anticoagulants, compounds more potent than their predecessors. One of these is brodifacoum, which is usually toxic in single doses. But death is delayed and therefore, as with other anticoagulants, no poison shyness develops in sublethally poisoned animals. Unlike the first generation anticoagulants, repeated feeding is not necessary for the poison to take effect. Pulsed baiting with many small baits at about weekly intervals minimizes social interaction effects of rodent hierarchies and maximizes the exposure of individual rodents. This technique is not possible with the earlier anticoagulants. Second generation anticoagulants are revolutionizing rodent control in agriculture, which is, for the first time, practical and effective.

Apart from the choice of toxicant, the timing of rodent control and the coordinated execution of a planned campaign are important in serious control programs. The most effective time to control agricultural rodents, for example, is when little food is available to them and when population is low (probably just before breeding). A few "avant-garde" methods of rodent control are sometimes suggested (eg chemosterilants, ultrasonic sound, electromagnetism) but none can yet claim to be as effective as anticoagulant rodenticides.   ACD

SQUIRREL-LIKE RODENTS

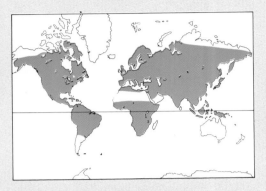

Suborder: Sciuromorpha
Seven families: 65 genera: 377 species.
Distribution: worldwide except for Australia region, Polynesia, Madagascar, southern S America, desert regions outside N America.

Habitat: very diverse, ranging from semiarid desert, flood plains, wetlands associated with streams, scrub, grassland, savanna to rain forest.

Size: head-body length from 2.2in (5.6cm) in the pocket mouse *Perognathus flavus* to 47in (120cm) in the beaver; weight from 0.35oz (10g) in the African pygmy squirrel and *Perognathus flavus* to 66lb (30kg) in the beaver.

Beavers
Family: Castoridae
Two species in 1 genus.

Mountain beaver
Family: Aplodontidae
One species.

Squirrels
Family: Sciuridae
Two hundred and sixty-seven species in 49 genera.

Pocket gophers
Family: Geomyidae
Thirty-four species in 5 genera.

Scaly-tailed squirrels
Family: Anomaluridae
Seven species in 3 genera.

SQUIRRELS are predominantly seedeaters and are the dominant arboreal rodents in many parts of the world, but in the same family there are almost as many terrestrial species, including the ground squirrels of the open grasslands, also mostly seedeaters, and the more specialized and herbivorous marmots.

Athough they may be highly specialized in other respects, members of the squirrel family have a relatively primitive, unspecialized arrangement of the jaw muscles and therefore of the associated parts of the skull, in contrast to the mouse-like ("myomorph") and cavy-like ("caviomorph") rodents which have these areas specialized in ways not found in any other mammals. In squirrels the deep masseter muscle is short and direct, extending up from the mandible to terminate on the zygomatic arch. This feature is shared by some smaller groups of rodents, notably the Mountain beaver (Aplodontidae), the true beavers (Castoridae), the pocket gophers (Geomyidae) and the pocket mice (Heteromyidae), and has led to these families all being grouped with the squirrels in a suborder, Sciuromorpha. However, retention of a primitive character is not by itself an indication of close relationship and these groups are so different in other respects that it is fairly generally agreed that the suborder is not a natural group.

These families appear to have diverged from each other and from other rodents very early in the evolution of rodents and have very little in common other than the retention of the "sciuromorph" condition of the chewing apparatus.

A further primitive feature retained by these rodents is the presence of one or two premolar teeth in each row, giving four or five cheekteeth in each row instead of three as in the murids. GBC

▶ **The profile of a squirrel:** a distinctive head, long cylindrical body and bushy tail.

▼ **Distinguishing feature of squirrel-like rodents.** The lateral masseter muscle (green) extends in front of the eye onto the snout, moving the lower jaw forward during gnawing. The deep masseter muscle (blue) is short and used only in closing the jaw. Shown here is the skull of a marmot.

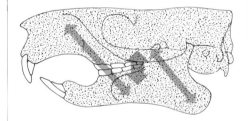

Pocket mice
Family: Heteromyidae
Sixty-five species in 5 genera.

Springhare or **Springhass**
Family: Pedetidae
One species.

Beaver

Red squirrel

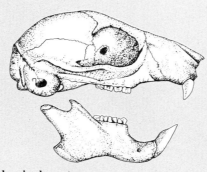

Red squirrel 1.7in

Skulls and teeth of squirrel-like rodents.

Skulls of squirrels show few extreme adaptations although those of the larger ground squirrels, such as the marmots, are more angular than those of the tree squirrels. Most members of the squirrel family have rather simple teeth, lacking the development of either strongly projecting cusps or of sharp enamel ridges as found in many other rodents. In the beavers, however, there is a pattern of ridges, adapted to their diet of bark and other fibrous and abrasive vegetation and convergent with that found in some unrelated but ecologically similar rodents, such as the coypu.

BEAVERS

Family: Castoridae
Two species in genus *Castor*.
Distribution: N America, Scandinavia, W and E Europe, C Asia, NW China.

Habitat: semiaquatic wetlands associated with ponds, lakes, rivers and streams.

Size: head-body length 32–47in (80–120cm); tail length 10–20in (25–50cm); shoulder height 12–23in (30–60cm); weight 24–66lb (11–30kg). No difference between sexes.

Coat: yellowish brown to almost black; reddish brown is most common.

Gestation: about 105 days.

Longevity: 10–15 years.

North American beaver
Castor canadensis
North American or Canadian beaver.

Distribution: N America from Alaska E to Labrador, S to N Florida and Tamaulipas (Mexico). Introduced to Europe and Asia.

European beaver
Castor fiber
European or Asiatic beaver.

Distribution: NW and C Eurasia, in isolated populations from France E to Lake Baikal and Mongolia.

▶ **At home in the water.** ABOVE Of all rodents the beaver is one of the best-adapted for movement in water—torpedo-shaped with waterproof fur, commanding thrust from its flattened tail and webbed feet.

▶ **Kits are precious:** a female beaver produces only one small litter per year. This well-developed kit is being carried by its parent's front feet and mouth.

Few wild animals have had as much influence on world exploration, history and economics as the North American beaver. Much of the exploration of the New World developed from the fur trade—stimulated principally by the demand for beaver fur for making the felt hats that were very popular in the late 18th and early 19th centuries. Even wars were fought over access to beaver trapping areas, such as the French and Indian War (or the Seven Years' War) of 1754–63 in which the British defeated the French and gained control of northern North America.

Biologically, the beaver's flat, scaly tail, webbed hind feet, huge incisor teeth and unique internal anatomy of the throat and digestive tract are all distinctive. Beavers display a rich variety of constructional and behavioral activities, perhaps unsurpassed among mammals. Their dams promote ecological diversity, affect water quality and yields and have affected landscape evolution. Beavers also have tremendous inherent interest (testified by the quantity of literature on them), partly because they display many human traits, such as family units, complex communication systems, homes (lodges and burrows), food storage, and transportation networks (ponds and canals).

Beavers are the second heaviest rodent in the world—occasionally attaining a weight of over 66lb (30kg). They are adapted for a semiaquatic life, with a torpedo-shaped body and large webbed hind feet. The beaver's large tail—horizontally flattened and scaly—provides both steering and power; it can be flexed up and down to produce a rapid burst of speed. When the beaver dives its nose and ears close tight and a translucent membrane covers the eyes. The throat can be blocked by the back of the tongue, and the lips can close behind the incisors so the animal can gnaw and carry sticks underwater without choking.

On land the beaver is slow and awkward: it waddles clumsily on its large, pigeon-toed rear feet and on its small, short front legs and feet. Its posture—nose down, pelvis up—creates an impression of a walking wedge. If alarmed on land it will gallop or hop towards water.

The North American beaver can be found across most of the North American continent (its historic range), but only because populations were reestablished by State and federal wildlife agencies. In the eastern USA the animal came close to extirpation in the late 19th century. Populations of the North American beaver have been introduced in

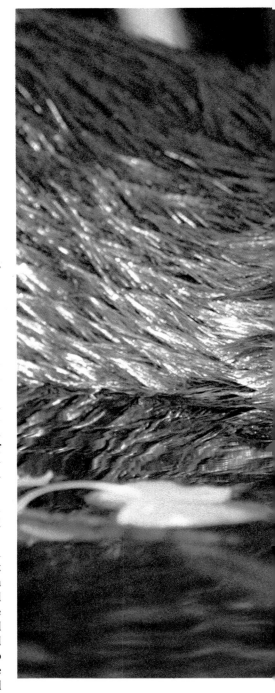

Finland, Russia (Karelian Isthmus, Amur basin and Kamchatka peninsula), and Poland.

The European beaver was once found throughout Eurasia but only isolated populations survive, in France (Rhone), Germany (Elbe) and Scandinavia, and in central Russia. The beavers in many of these groups are classified as subspecies, but some scientists consider the west European beaver a distinct species, *Castor albicus*, mainly because of differences in its skull.

The earliest direct ancestor of Eurasian beavers was probably *Steneofiber* from the Middle Oligocene (about 32 million years ago). The genus *Castor*, which originated in Europe during the Pliocene (7–3 million years ago), entered North America and because of its geographic isolation evolved into the present species. Thus the North

American beaver is considered to be younger and more advanced than the European beaver. During the Pleistocene, 10,000 years ago, these two species co-existed with giant forms which weighed 600–700lb (270–320kg), for example *Castoroides* in North America.

Beavers are herbivores whose diet varies with changes in season. In spring and summer they feed on nonwoody plants or plant parts, eg leaves, herbs, ferns, grasses and algae. In fall they favor woody items, taking them from a great variety of tree species but preferring aspen and willow. The digestion of woody plant material and cellulose is enhanced by microbial fermentation in the cecum and by the reingestion of the cecal contents.

Thanks to the adaptations of its teeth the beaver possesses a remarkable ability to cut down trees, for food and building materials. Like the incisor teeth of all living rodents the beaver's incisors are large and grow as fast as they are worn down by gnawing. They have a tough outer layer and a softer inner layer, forming a chisel-like edge which is kept sharp by rubbing or gnashing. Through them the beaver can exert enormous pressure when cutting.

During the fall, in northern climates, beavers gather woody stems as food for the winter. These are stored underwater, near a beaver's winter lodge or burrow. The water acts as a refrigerator, keeping the stems at a temperature around (32°F) 0°C, preserving their nutritional value.

Beavers live in small, closed family units (often incorrectly termed "colonies") which usually consist of an adult pair and offspring from one year or more. An established family contains an adult pair, kits of the current year (up to 12 months old), yearlings born the previous year (12–24 months old), and possibly one or more subadults of either sex from previous breeding seasons (24 or more months old). The subadults generally do not breed.

The beaver's social system is unique among rodents. Each family occupies a discrete, individual territory. In northern latitudes male and female have a "monogamous" and long-term relationship. Their family life is exceptionally stable, being based on a low birth rate (one litter per year of 1–5 young in the European beaver and up to eight in the North American beaver), a high survival rate, parental care of a high standard and up to two years in the family for the development of adult behavior. The family is structured by a hierarchy in which adults dominate yearlings and yearlings dominate kits, each conveying its own status by means of vocalizations, postures and gestures. Physical aggression is rare. Mating occurs during winter, in the water. Kits are born in the family lodge in late spring, soon after the yearlings have dispersed for distances usually less than 12mi (20km) but occasionally up to 155mi (250km). At birth they have a full coat of fur, open eyes and are able to move around inside the lodge. Within a few hours they are able to swim, but their small size and dense fur makes them too buoyant to submerge easily, so they are unable to swim down the passage from the lodge. Kits nurse for about 6 weeks and all members of the family share in bringing solid food to them with the male most actively supplying provisions. The kits grow rapidly but require many months of practice to perfect their ability to construct dams and lodges.

One of the ways by which beavers communicate is by depositing scent, often around the edges of water areas occupied by a family. The North American beaver scent marks on small mounds made of material dredged up from under water and placed on the bank whereas the European beaver marks directly on the ground. The scent, produced by castoreum from the castor sacs and secretions from the anal glands, is pungent and musty. All members of a beaver family participate, but the adult male marks most frequently. Scent marking is most intense in the spring and probably conveys information about the resident family to dispersing individuals and to adjacent families.

Beavers also communicate by striking their tails against the water (tail-slapping). This is done more often by adults than by kits, usually when they detect unusual stimuli. The slap is a warning to other members of the family who will move quickly to deep water. It may also frighten enemies.

Beaver construction activities modify their environment to provide greater security from predators, enhance environmental stability, and permit more efficient exploitation of food resources. Of the various construction activities in which beavers engage canal building is the least complex and was probably the first that beavers developed. Beavers use their forepaws to loosen mud and sediment from the bottom of shallow streams and marshy trails and push it to the side out of the way. Repeated digging and pushing of material creates canals that enable beavers to remain in water while moving between ponds or to feeding areas. This behavior occurs most frequently in summer when water levels are low and all members of the family participate.

Dams are built across streams to impound water, using mud, stones, sticks and branches. In northern climates the ponds created make the beavers' lodges more secure from predators and permit them to use more distant and larger food items. They must also be deep enough for the family to be able to swim under the ice from their lodge to the food cache. The sound of flowing water and visual cues stimulate dam building. Beavers use their small, agile forepaws to loosen mud, small stones and earth from the stream bottom and carry the material to the dam site. Sticks and branches are towed with the incisors to the dam. The branches support the dam and keep the mud

► **The towering jumble** of a beaver's dam—6 to 9 feet high—conceals the careful, firm construction of mud, stones, sticks and branches.

► **Silent approach.** BELOW A beaver tows a branch to the base of its dam.

▼ **The beaver's lodge** or house is a large conical pile of mud and sticks, either on the shore or in the middle of a pond. The structure begins with beavers digging a burrow underwater into the bank (or into a slight mound adjacent to a waterway) with their forepaws. The burrow is extended upwards towards the surface as the water level rises behind the dam, and if it breaks through the ground the exposed hole is covered with sticks, branches and mud. Material continues to be added on top of the burrow while the living chamber is hollowed out above the water and within the mound of sticks and mud. In many cases the rising water isolates the lodge in the middle of the beaver pond. Most lodges are begun in later summer or early fall. Each fall in northern latitudes families add a thick coating of mud and sticks to the exterior of the lodge in which the family will spend the winter. Frozen mud provides insulation and prevents predators from digging into the lodge. All family members, except kits, help to build and maintain the lodges. Females are more active than males and the adult female is the most active lodge builder.

The Beaver's 29-hour Day

During most of the year beavers are active at night, rising at sunset and retiring to their lodges at sunrise. This regular daily cycle of activity is termed a circadian rhythm. In northern winters, however, where ponds freeze over, beavers stay in their lodges or under the ice because temperatures there remain near 32°F (0°C) while air temperatures are generally much lower. Activity above ice would require a very high production of energy.

In the dimly lit water world of the lodges and the surrounding water, light levels remain constant and low throughout the 24-hour day, so that sunrise and sunset are not apparent. In the absence of solar "cues" activity, recorded as noise and movement, is not synchronized with the solar day above. The circadian rhythm breaks down and so-called beaver days become longer. This may be an incidental consequence of the beaver's ability to live under the ice, but the biological significance of this change in activity is not clearly known. The beaver days vary in length from 26 to 29 hours, so during the winter beavers experience fewer "days" under the ice than occur above. This type of cycle is termed a free-running circadian rhythm.

and small sticks in place. Beavers keep adding mud and sticks to make the dam higher and longer; some may reach over 330ft (100m) long and 10ft (3m) high. Dam building activity is most intense during periods of high water, primarily spring and fall, although material may be added throughout the year. Adults and yearlings build dams, but female members of the family are more active than males and the adult female is most active.

Beavers are harvested for their fur and meat and are actively managed throughout most of their range. Annual harvests in Canada since the 1950s amounted to 200,000 to 600,000 pelts and during the 1970s in the USA 100,000 to 200,000. In Eurasia recent harvests from Nordic countries took less than 1,000 pelts per year and in Russia about 8,500 in 1980.   RAL/HEH

MOUNTAIN BEAVER

Aplodontia rufa
Sole member of family Aplodontidae.
Mountain beaver or beaver or boomer or
sewellel.
Distribution: USA and Canada along Pacific
coast.

Habitat: coniferous forest.

Size: head-body length 12–16in
(30–41cm); tail length 1–1.5in
(2.5–3.8cm); shoulder height
4.5–5.5in (11.5–14cm); weight
2.2–3.3lb (1–1.5kg). Sexes are
similar in size and shape, but
males weigh slightly more than
females.

Coat: young in their year of birth have grayish
fur, adults blackish to reddish brown, tawny
underneath.

Gestation: 28–30 days.

Longevity: 5–10 years.

Mountain beavers, the most primitive of all present-day rodents, live only in southwest Canada and along the west coast of the USA, where they inhabit some of the most productive coniferous forest lands in North America. Not to be confused with the flat-tailed stream beaver (*Castor*), Mountain beavers are land animals that spend much of their time in burrows, where they sleep, eat, defecate, fight and reproduce and do most of their traveling. They are seldom seen outside zoos.

Mountain beavers are medium-sized, bull-necked rodents with a round, robust body and a moderately flat and broad head with small black, beady eyes and long, stiff, whitish whiskers (vibrissae). Not only are their incisor teeth rootless and continuously growing but also their premolars and molars. Their ears are relatively small, covered with short, soft, light-colored hair. Their short legs give them a squat appearance; a short vestigial tail makes them look tailless. The fur on their back presents a sheen whereas white-to-translucent-tipped long guard hairs give their flanks a grizzled affect. A distinctive feature is a soft, furry white spot under each ear.

Mountain beavers can be found at all elevations from sea level to the upper limit of trees and in areas with rainfall of 20–138in (50–350cm) per year, and where winters are wet, mild and snow-free, summers moist, mild and cloudy. Their home burrows are generally in areas with deep, well-drained soils and abundant fleshy and woody plants. In drier areas Mountain beavers are restricted to habitats on banks and to ditches that are seasonally wet or have some free-running water available for most of the year.

The location of Mountain beavers is in part explained by the fact that they cannot adequately regulate the temperature of their bodies and must therefore live in stable, cool, moist environments. Nor can they effectively conserve body moisture or fat, which prevents them from hibernating or spending the summer in torpor. They need to consume considerable amounts of water and food and to line their nests well for insulation. They can satisfy most of their requirements with items obtained within 100ft (about 30m) of their nest. Water is obtained mainly from succulent plants, dew or rain.

When defecating, Mountain beavers, unlike any other rodent or rabbit, extract fecal pellets individually with their incisors and toss them on piles in underground fecal chambers. Like rodents and rabbits they also reingest some of the pellets.

Mountain beavers are strictly vegetarians. They harvest leafy materials (such as fronds of Sword fern, new branches of salal and huckleberry, stems of Douglas fir and vine maple, and clumps of grass or sedge) and succulent, fleshy foods (such as fiddle heads of Bracken fern, roots of False dandelion and bleeding heart). These are eaten immediately or stored underground.

Most food and nest items are gathered above ground between dusk and dawn and consumed underground. Decaying uneaten food is abandoned or buried in blind chambers; dry, uneaten food is added to the nest.

It used to be thought that all Mountain beavers were solitary animals, except during the breeding and rearing season. Recent studies using radio tracking have demonstrated that this view is inadequate. Some Mountain beavers spend short periods together, in all seasons. Neighbors, for example, will share nests and food caches, or a wandering beaver will stay a day or two in another beaver's burrow system. Sometimes a beaver's burrow will also be occupied, in part if not in its entirety, by other animals, eg salamanders, frogs or deer mice.

Unlike many rodents, Mountain beavers have a low rate of reproduction. Most do not mate until they are at least two years old and females conceive only once a year, even

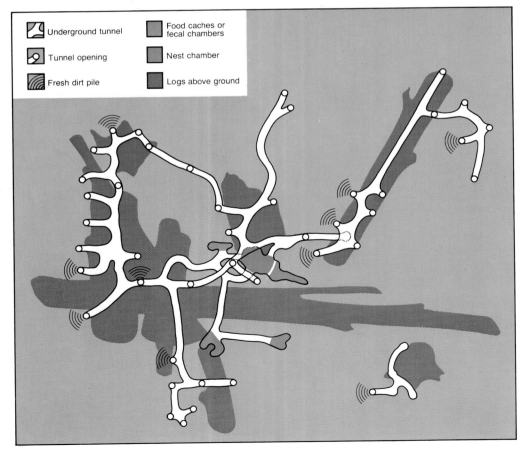

Underground tunnel
Tunnel opening
Fresh dirt pile
Food caches or fecal chambers
Nest chamber
Logs above ground

▲ **Out foraging.** Mountain beavers rarely
appear above ground in daylight. This
exceptional picture, showing well the contrast
between the sheen of the back and the grizzled
flanks, offers a portrait of the Mountain beaver:
plump-bodied, bull-necked, snouty.

◀ **The burrow system of a Mountain beaver.**
Each consists of a single nest chamber and
underground food caches and fecal chambers
which are generally close to the nest. Most
nests are about 3.3ft (1m) below ground in a
well-drained, dome-shaped chamber, although
some may occur 6.6ft (2m) or more below
ground. Tunnels are generally 6–8in
(15–20cm) in diameter and occur at various
levels; those closest to the surface are used for
travel; deep ones lead to the nest and food
caches.

Burrow openings occur every 13–20ft
(4–6m) or more depending on vegetative cover
and number of animals occupying a particular
area. Densities vary from 1 Mountain beaver
per acre (0.4ha) in poor habitat to 8 or more
per acre in good habitat. Up to 30 per acre have
been kill-trapped in reforested clear-out areas,
but such densities are rare. Individual systems
often connect.

if they lose a litter. Breeding is to some extent
synchronized. Males are sexually active
from about late December to early March,
with aggressive older males doing most of
the breeding. Conception normally occurs
in January or February. Litters of 2–4 young
are born in February, March or early April.
Young are born blind, hairless and helpless
in the nest. They are weaned when about
6–8 weeks old and continue to occupy the
nest and burrow system with their mother
until late summer or early fall. Then they
are forced out, to seek sites for their own
homes.

Once on its own a young beaver may
establish a new burrow system or, more
commonly, restore an abandoned one. New
burrows may be within 330ft (100m) of the
mother's burrow or up to 1.2mi (2km)
away. The distance depends on population
densities and on the quality of the land. Both
sexes disperse in the same manner.

Dispersing young travel mainly above
ground: they become very vulnerable to
predators, for example owls, hawks and

ravens, coyotes, bobcats and man. Those
living near roads are liable to be killed by
vehicles.

Mountain beavers are classed as non-
game mammals, hence they are not managed like game (eg deer, elk) or like furbear-
ing animals (eg beavers, muskrat). The
damage they cause affects about 275,000
acres (111,000ha) of forest; they pose a tremendous threat to young conifers planted
for timber. Almost all damage occurs while
Mountain beavers are gathering food and
nest materials. The result of their activities is
the loss of timber worth several million US
dollars every year.

Although some of the Mountain beavers'
forest habitat has given way to urban
development and agriculture, they range
over about as extensive an area now as they
did 200 to 300 years ago. They are probably
more abundant now than they were in the
early 20th century thanks to forest logging
practices. Mountain beavers do not appear
to be in any immediate danger of extermin-
ation from man or natural causes. JE

SQUIRRELS

Family: Sciuridae
Two hundred and sixty-seven species in 49 genera.
Distribution: worldwide apart from the Australian region and Polynesia, Madagascar, southern S America, desert regions (eg Sahara, Egypt, Arabia).

Habitat: varies from lush tropical rain forest to city parks and semiarid desert.

Size: ranges from head-body length 2.6–3.9in (6.6–10cm), tail length 2–3.1in (5–8cm), weight about 0.35oz (10g) in the African pygmy squirrel to head-body length 20.8–28.7in (53–73cm), tail length 5.1–6.3in (13–16cm), weight 8.8–17.6lb (4–8kg) in the Alpine marmot.

Gestation: about 40 days (ground squirrels and sousliks 21–28 days; marmots 33 days).

Longevity: varies considerably to maximum of 8–10 years (up to 16 in captivity).

Species and genera include: **African pygmy squirrel** (*Myosciurus pumilio*), **Alpine marmot** (*Marmota marmota*), **American red squirrel** (*Tamiasciurus hudsonicus*), **Arctic ground squirrel** (*Spermophilus undulatus*), **beautiful squirrels** (genus *Callosciurus*), **Black giant squirrel** (*Ratufa bicolor*), **chipmunks** (genus *Tamias*), **Cream giant squirrel** (*Ratufa affinis*), **Douglas pine squirrel** (*Tamiasciurus douglasii*), **flying squirrels** (genera *Aeromys, Belomys, Eupetaurus, Glaucomys, Hylopetes, Petaurista, Petinomys, Pteromys, Pteromyscus, Trogopterus*), **Fox squirrel** (*Sciurus niger*), **giant flying squirrels** (genus *Petaurista*), **giant squirrels** (genus *Ratufa*), **Gray squirrel** (*Sciurus carolinensis*), **ground squirrels** (genus *Spermophilus*), **Horse-tailed squirrel** (*Sundasciurus hippurus*), **Kaibab squirrel** (*Sciurus kaibabensis*), **Little souslik** (*Spermophilus pygmaeus*), **Long-clawed ground squirrel** (*Spermophilopsis leptodactylus*), **Long-nosed squirrel** (*Rhinosciurus laticaudatus*), **marmots** (genus *Marmota*), **Plantain squirrel** (*Callosciurus notatus*), **prairie dogs** (genus *Cynomys*), **Prevost's squirrel** (*Callosciurus prevosti*), **Red squirrel** (*Sciurus vulgaris*), **Russian flying squirrel** (*Pteromys volans*), **Siberian chipmunk** (*Tamias sibiricus*), **Slender squirrel** (*Sundasciurus tenuis*), **Southern flying squirrel** (*Glaucomys volans*), **Spotted giant flying squirrel** (*Petaurista Elegans*), **Sunda squirrels** (genus *Sundasciurus*), **Three-striped ground squirrel** (*Lariscus insignis*), **tree squirrels** (genus *Sciurus*), **woodchuck** (*Marmota monax*).

LIKE Squirrel Nutkin in Beatrix Potter's children's story *The Tale of Squirrel Nutkin*, squirrels have a popular image as attractive but cheeky opportunists. Their opportunism, however, is the basis of their success. Today squirrels are among the most widespread of mammals, and are found on every continent other than Australia and Antarctica. Because they are relatively unspecialized, squirrels have been able to evolve a wide variety of body forms and habits which fit them for life in a broad range of habitats, from lush tropical rain forests to rocky cliffs or semiarid deserts, from open prairies to town gardens. This successful family includes such varied forms as the ground-dwelling and burrowing marmots, ground squirrels, prairie dogs and chipmunks; the arboreal and day-active tree squirrels; and nocturnal flying squirrels.

Squirrels range in size from the tiny, mouse-like and arboreal African pygmy squirrel to the sturdy cat-sized marmots. Typically squirrels have a long, cylindrical body with a short or long bushy tail. Most squirrels have fine soft hair and in some species the coat is very thick and valued by man as fur. Many species molt twice a year and often sport a summer coat lighter in color. Male, female and young are similar in appearance, but even within species there can be considerable variation in color as for instance in the variable squirrels of Thailand where some populations are pure white, others pure black or red and yet others a combination of these and other colors.

Squirrels usually have large eyes, and tree and flying squirrels have large ears. Some species, eg the European red squirrel and the Kaibab squirrel, have conspicuous ear tufts. They have the usual arrangement of teeth in rodents, viz a single pair of chisel-shaped incisor teeth in each jaw and a large gap in front of the premolars because canines are absent. The incisors grow continuously and are worn back by use; the cheek teeth are rooted and have abrasive chewing surfaces. The lower jaw is quite movable, and some genera like chipmunks and ground squirrels have cheek pouches for carrying food.

Squirrels have sharp eyesight and wide vision but the ability of flying squirrels to see color is poor. In European red squirrels, marmots and prairie dogs the entire retina has the sensitivity that most mammals possess only within a small area of the retina (the foveal region). These squirrels can distinguish vertical objects particularly well, an ability important for tree-dwelling species, which are able to estimate distances between trees with great accuracy. Many

species also have a well-developed sense of touch, thanks partly to touch-sensitive whiskers (vibrissae) on the head, feet and the outside of the legs.

Squirrels have short forelimbs, with a small thumb and four toes on the front feet, and longer hindlimbs, with four (woodchuck) or five toes on the hind feet (tree squirrels). Although the "thumb" is usually poorly developed it can have a very long claw (prairie dogs) or flat nail (marmots); all other toes have sharp claws. The soles of the feet have soft pads; in the desert-living long-clawed ground squirrels they are covered in hair which acts as an insulator to enable the animal to move over hot sand. These squirrels also have fringes of stiff hairs on the outside of the hind feet which push away the soft sand as they burrow. Among other adaptations are those of several of the terrestrial squirrels which nest and take refuge in underground burrows: they are heavy bodied with powerful forelimbs and large scraping claws for digging.

▶ **Pinned to the bark.** Tree squirrels, such as this Arizona gray squirrel, hold themselves head-down while waiting to make the next move.

▼ **A shrill screeching** means that this Long-tailed marmot has been disturbed or feels threatened.

Tree squirrels live and make their nests (dreys) in trees and are excellent at climbing and jumping; their toes, with their sharp claws, are well adapted for clinging to tree trunks. Squirrels descend tree trunks head first, sticking the claws of the outstretched hind feet into the bark to act as anchors. The tail serves as a balance when the squirrel runs and climbs and as a rudder when it jumps. The tail is also used as a flag to communicate social signals and is wrapped around the squirrel when the animal sleeps. On the ground, tree squirrels move in a sequence of graceful leaps, often pausing between jumps to raise their heads and look around. When feeding, all squirrels squat on their haunches holding the food between their front paws.

Tree squirrels are able to jump across the considerable gaps in the canopy. In a flying leap the legs are outspread and the body flattened, with the tail slightly curved so that a broad body surface is presented to the air. It is only a short step from such a leap to the controlled glides of the true "flying" squirrels. Like the other gliding mammals (the flying lemurs and flying phalangers) these squirrels have developed a furred flight membrane or patagium which extends along the sides of the body and acts like a parachute when the animal leaps. It extends from the hind legs to the front limbs and is bound in front by a thin rod of cartilage attached to the wrist. The bushy tail is free and acts like a rudder. The squirrel steers by changing the position of the limbs and tail and the tension in the muscular gliding membrane. Flying squirrels descend in long smooth curves, avoiding branches and trees, to land low on a tree trunk. As it lands the flying squirrel brakes, by turning its tail and body upward. The larger flying squirrels can glide for 330ft (100m) or more but the smaller forms cover much shorter distances. Gliding is an economical way to travel through the tall forest and enables the squirrel to escape quickly from a predator such as a marten. The flight membrane does have some drawbacks, however; when moving around in trees flying squirrels are less agile than tree squirrels and it is probably significant that the mammals that have adopted a gliding habit are active only at night when they are less conspicuous to keen-sighted birds of prey.

Most squirrels feed on nuts and seeds, fruits and other plant material, supplemented with a few insects, though they have been known to feed on reptiles and young birds. At certain times of year fungi and insects may be eaten in quantity. For grow-

ing juvenile Gray squirrels in English woodland insects are an important source of protein in early summer when other nutritious foods are scarce. Several tropical squirrels have become totally insectivorous and the incisors of the long-nosed squirrel *Rhinosciurus laticaudatus* have become modified into forceps-like structures for grasping the insects that make up the animal's diet.

The Gray squirrel in England regularly strips bark (see pp42–43) and many other species will do so if food is short. Squirrels of the genus *Sundasciurus* specialize in feeding on bark and sap from the boles of trees in Malayan rainforest. Many species take large quantities of leaf matter, especially young leaves; the Giant flying squirrel is primarily a leaf-eater. Red squirrels feast on young conifer shoots, and thereby cause much damage in forestry plantations. Terrestrial squirrels feed mainly on low-growing plants and may affect the vegetation where they live. Prairie dogs feed in the immediate area around their burrows; they bite off all tall plants to open up a field of view. They feed on herbs and grasses—cropping plants and keeping vegetation low. The process induces changes in vegetation, encouraging the development of fast-growing plants. Prairie

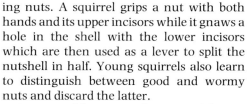

▲ **Bushy-tailed mountain rodent.** The heavy Hoary marmot dwells in the mountainous areas of northern Canada and Alaska. Reserves of fat accumulated during summer for hibernation can amount to 20 percent of body weight.

◄ **Tucked away,** the gliding membrane of the Southern flying squirrel does not impede the animal's movement on tree trunks.

▼ **Membrane full spread,** a Giant flying squirrel glides between trees at night.

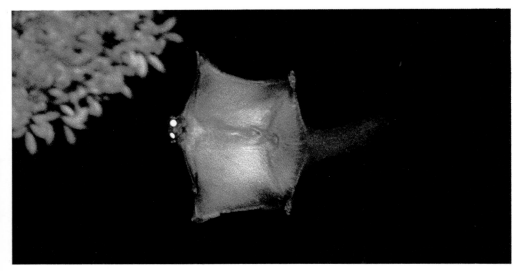

ing nuts. A squirrel grips a nut with both hands and its upper incisors while it gnaws a hole in the shell with the lower incisors which are then used as a lever to split the nutshell in half. Young squirrels also learn to distinguish between good and wormy nuts and discard the latter.

Because of the seasonal nature of flowering and fruiting in temperate forests the squirrels that live there must depend on different foods in different seasons but nuts and seeds are often cached, for use during the winter. From July, when the first fruits mature, until the following March Gray squirrels in English woodlands for instance depend on fresh and buried hazelnuts, acorns and beech mast. A poor mast crop can have serious consequences for squirrel populations, possibly preventing breeding the following spring or provoking large-scale migrations, over several miles, usually of younger subordinate animals. Even infant squirrels hoard and bury food and show instinctive innate burying behavior. Adult Red squirrels can smell and relocate pine cones buried 12in (30cm) below the surface. A squirrel does not necessarily find the hoard it buried—it might find another's. Many hoards are never retrieved and hence the burying of surplus food disperses seeds in temperate forests. This is not the case in tropical forests where the role of squirrels is more that of seed predators. There has been a long coevolution of squirrels with many cone- and mast-producing trees (eg pine, hazel, oak, beech, chestnut). Because their dentition enables them to gnaw through hard tissue and destroy the nutritive fruit embryo, squirrels are only successful seed dispersers where foods are seasonally superabundant and the animals make food caches. Gray squirrels carry acorns up to 100ft (about 30m) from a fruiting tree and bury them. Since many of these buried acorns are not found again they survive to germinate and eventually become new oak trees.

The Douglas pine squirrel has an equally important role for several conifers as it cuts unopened pine cones and caches them in damp places under logs or in hollow stumps in stores of 160 cones or more.

Many ground squirrels hibernate during the harsh winter months and prepare for this hibernation by laying up food supplies in their dens (tree squirrels do not hibernate though they may not emerge from their nests for a few days in bad weather). The Siberian chipmunk can carry up to 0.3oz (9g) of grain in its cheek pouches for a distance of over 0.6mi (1km). It stores seeds,

dogs and most squirrels get all the water they require from the plant material they ingest but the European red squirrel in particular must stay close to sources of drinking water. Trappers take advantage of this fact and use water to attract squirrels into traps during a hot summer.

For most tree squirrels nuts and seeds are a major and highly nutritious part of the diet. One Red squirrel in Siberia took 190 pine cones in one day, eating the seeds and discarding the winged part. Squirrels have a special innate levering technique for open-

buds, acorns and mushrooms all stashed in different compartments adjoining the sleeping quarters of its underground burrow. One animal may store as much as 4.5–13lb (2–6kg) of food for its winter supplies. Similarly ground squirrels and sousliks become quite fat before hibernation, which lasts 5–7 months, but also store winter food supplies which are generally used only in spring when the animals reawaken. Marmots are also true hibernators. The entire marmot family unit (up to 15 animals) retreats into its den and sleeps huddled together throughout the winter. The last animal in, usually an adult male, plugs the entrance hole from the inside with hay, earth and stone. Marmots hibernate for 6 months or so and while they sleep their metabolism slows down and their heartbeat, body temperature and respiratory rate drop. When the temperature outside is below freezing the hibernating marmots have a body temperature of 40–45.5°F (4.5–7.5°C). Every 3–4 weeks the marmots wake to defecate and urinate and are much slimmer when they emerge in the spring, when they promptly begin to spring clean their burrows.

Some ground squirrels, eg the Arctic ground squirrel, the Little souslik and the Long-clawed ground squirrel, not only hibernate during the winter but also sleep during part of the summer when there is drought and the vegetation has withered away. They close off their dens with grass and sand and then settle down to sleep.

Most squirrels are sexually mature and able to breed when one year old, though marmots are not fully grown until they are two. Hibernating species usually mate a few days after the end of hibernation; marmots mate in May while still in their winter burrows and sexual activity is stimulated by odors from the anal glands of both sexes. Marmot families consist of both males and females—males are tolerant of each other even during the mating season. Gestation lasts about five weeks and the pregnant female closes off her living quarters with hay several days before the birth. Like all infant squirrels baby marmots are born naked, toothless and helpless with their eyes closed, but by six weeks of age they are fully-furred and sufficiently independent to be able to venture out of the burrow. While the young marmots play, other members of the family stand guard. In prairie dog societies too, both sexes are friendly towards the pups and take responsibility for them. The young

◀ ▲ **Representative species of squirrels.**
(**1**) A Southern flying squirrel (*Glaucomys volans*) gliding from a nest hole in a tree trunk.
(**2**) Prevost's squirrel (*Callosciurus prevosti*).
(**3**) African pygmy squirrel (*Myosciurus pumilio*) descending a tree head first. (**4**) An Abert or Tassel-eared squirrel (*Sciurus aberti*). (**5**) An American red squirrel (*Tamiasciurus hudsonicus*) hanging by its hindlegs. (**6**) An Indian giant or Malabar squirrel (*Ratufa indica*). (**7**) An Asiatic or Siberian chipmunk (*Tasmias sibiricus*) with filled cheek pouches. (**8**) Alpine marmot (*Marmota marmota*) in vigilant upright posture giving alarm whistle. (**9**) A Shrew-faced ground or Long-nosed squirrel (*Rhinosciurus laticaudatus*) foraging for termites. (**10**) Geoffrey's or Western ground squirrel (*Xerus erythropus*) with its tail arched and fluffed, an indication of anxiety.

marmots hibernate and live together in the parents' burrow for the next summer.

The situation described above where both male and female help care for the young is exceptional among squirrels; in most species parental care is the sole responsibility of the female. In spring male woodchucks and Red and Gray squirrels may travel considerable distances to mate. Gray and Red squirrels have mating chases with several males following a receptive female. After mating the male usually has no further association with the female who lines the breeding drey or den and rears the young alone. If she feels the young are threatened the mother moves them to another nest, carrying them by the scruff of the neck.

In all squirrel species gestation is short: 3–6 weeks. The young are always born naked, toothless and helpless with eyes closed. Litter sizes range from one or two (Spotted flying squirrel, Giant squirrel) to as many as nine or more (Gray squirrel); litter sizes may vary according to the age and condition of the mother. Many species breed only once a year (prairie dogs) but Gray and Red squirrels have two breeding seasons— in spring and summer—if conditions are favorable. If there has been little or no spring breeding because of winter food shortages large numbers of litters are produced in the summer to compensate.

The development of a young Red squirrel is fairly typical. At birth it will weigh only 0.28–0.42oz (8–12g) but it grows fast. Hair begins to appear at 10–13 days and it will be fully covered in hair by three weeks. The lower incisors appear at 22 days, the upper ones at 35. The eyes open after 30 days and the animal soon becomes more mobile and body cleaning and growing movements develop. It leaves the nest for the first time at 45 days, at which time it also takes its first solid food, and after 8 weeks is fully weaned and independent, though it remains near the mother and may still share her nest. Juvenile mortality is high with only about 25 percent of the young surviving to be one year old. Many fall to birds of prey or pine martens or die during migrations away from the maternal home range.

Development in young flying squirrels is slower than for many other mammals of similar size (eg cavies whose young are born fully furred and active); this is probably because of the hazards of nighttime arboreal activity and gliding. At birth infant flying squirrels already have a well-developed gliding membrane. Baby Southern flying squirrels are nursed for 60 or 70 days and are not active until that age.

Many of the terrestrial squirrels are very social animals. Alpine marmots live together in colonies ranging from 2 or 3 to 50 or more animals and one large colony may occupy an immense burrow system. Within the colony they scent mark their territory with substances secreted from the cheek glands and any intruder will be chased away by colony residents who gnash their teeth and call loudly. Once marmots have settled down and built their burrows they do not

▲ **The Red squirrel**–the archetypal squirrel. This species inhabits woodlands across Europe and Central Asia, but is hunted for its fur in the USSR.

▲ **The first summer** of a young prairie dog's life is a time of friendly relations with parents.

▷ **Half a prairie dog's life** (up to 15 months) OVERLEAF can be absorbed in growing to adult size.

▼ **The Black-tailed prairie dog** lives in colonies on plains and plateaus from southern Saskatchewan south to northern Mexico.

move far away from their home; in the mountains a family of marmots works about 0.6 acre (0.25ha).

Ground squirrels and sousliks also live socially in colonies with massive tunnel systems but some, like the European souslik, live singly within the colonies, each animal in its own den. Prairie dogs have an even more highly organized society: they live in social groups called coteries and several coteries make up "towns" which may cover areas of up to 160 acres (65ha). These interconnected burrows provide the prairie dogs with a refuge and a place to rear their young. Burrows may be surrounded by

crater-shaped mounds of earth, which help to prevent them flooding.

All animals from a coterie may use all parts of the coterie's burrow system. When coterie members meet they often touch noses; this is probably a form of identification or greeting. The dominant male usually maintains the coterie boundaries by challenging (with chases and calls) other prairie dogs from neighboring coteries. He may be helped by a subordinate coterie member, often his mate, who vocalizes at the intruder. Non-breeding coteries may consist of several males and females; one male usually dominates the other males but there is no clear dominance hierarchy among the females. In breeding coteries there is usually one adult male for every four adult females. Both sexes are friendly towards the young in their first summer and pups follow adults and play with them.

Relationships between coteries change with the season. In summer coterie boundaries are relaxed, and friendly contacts with neighboring coteries are common. By the fall coteries are exclusive and by December and January the dominant male is busy defending his boundaries against the neighbors. During the spring boundaries are gradually relaxed again, permitting cross-coterie contacts and inter-breeding. When

conditions become crowded adult prairie dogs emigrate, expanding the town's boundaries and leaving the old burrows to their offspring. This has good survival value in that older, more experienced animals leave to colonize new areas leaving the inexperienced young in familiar surroundings.

Something similar happens among American red squirrels where the female gives up her territory to her offspring. Although many Gray squirrel young emigrate from the area where they were born into areas of low population (if space allows), some establish home ranges within the original range of their mother, ie in territory with which they are already familiar and where they are likely to have a higher chance of survival. Gray squirrels are not territorial. Although they usually forage independently several squirrels will use the same area of woodland and often sleep together in the same dreys. Squirrels have overlapping home ranges with the male's ranges (10–15 acres, 4–6ha) larger than those of females (5 acres, 2ha) and one male's range will overlap those of several females.

There is a definite dominance hierarchy in Gray squirrels including both males and females; animals of low status are often forced to range widely and emigrate at times of food shortage. Animals taking part in these migrations are often the young of the year. It is these low status animals which cause the most damage in young plantations where they strip bark (see pp42–43).

In fact the Gray squirrel pattern of social organization with larger male ranges overlapping smaller female ranges is typical of non-territorial squirrels and of many other arboreal mammals. Individual Gray squirrels do not have the social advantages of mutually cooperative food searching or protection against predators in the form of an efficient alarm system and cooperative resistance (compare prairie dogs). Instead they are widely spread over the available food resources in such a way that competition is minimal but each squirrel has social contact with its neighbors either directly or by scent marking trees (by gnawing under branches, roots, etc and depositing urine) and via them with the whole population. Such a social system allows efficient outbreeding with plenty of social feedback to influence population changes such as timing of reproduction, range shift and emigration. Little is known about social organization in flying squirrels though some are

reported to sleep in pairs in tree holes, but they too probably have overlapping home ranges.

Squirrel population densities vary considerably according to the species, habitat and number of species present. Red squirrels in pine forests in Scotland occur at densities of only 0.3 animals per acre (0.8 animals per ha) whereas Gray squirrels often reach densities of 2 animals per acre (about 5 or 6 animals per ha) in deciduous woodlands in England. Total squirrel densities in tropical forests are usually much lower (less than 0.8 per acre or 2 animals per ha) in spite of the great diversity of species and the year-round productivity of the forest.

Factors that limit the density of day-active squirrels are the spacing and timing of available food and competition for these sources. One limiting factor for flying squirrels must be the availability of suitable tree holes where they rest during the day. The low densities of tropical squirrels can be explained by competition among squirrels and the high densities of other arboreal mammals, especially primates, which not only compete for the same foods but also have a long-term influence on the ecology and evolution of the forest.

The main natural predators of squirrel populations are carnivores such as weasels, foxes, coyotes, bobcats, martens and birds of prey and owls, but also man, who kills squirrels because they are agricultural pests, or for sport or for fur. Some species may be threatened by habitat loss due to timber felling or change of land use.

Man, however, has had a long association with and affection for many species of squirrel. Red squirrels were kept as pets by Roman ladies, marmots were taught to perform at medieval fairs and prairie dogs and woodchucks feature in American Indian legends.

Red squirrels are prominent in many folk stories and fairy tales and were particularly important in Indian and Germanic religions and myths. The Red squirrel was holy to the Germanic god Donar because of its color and in Germany and England squirrels were once sacrificed at the feasts of spring and the winter solstice. In an Indian saga squirrel dries up the ocean with its tail.

In the United States Gray and Fox squirrels are prized as game animals though elsewhere squirrels are shot on account of the damage they cause in young tree plantations. The burrowing activity of ground squirrels and sousliks can improve the land. But by bringing up soil from lower depths they may be a nuisance in agricultural

Niche Separation in Tropical Tree Squirrels

For two or more species of mammal to live in the same habitat their use of resources must be sufficiently different to avoid the competitive exclusion of one species by another. Such differences in life-style as ground-living or tree-dwelling, active in daytime or nocturnal, insect-eating or fruit-eating are obvious means of ecological separation between squirrel species. Often, however, species occurring naturally in the same habitat appear to be utilizing the same food resources but closer study reveals that each occupies a somewhat different niche.

The situation is well illustrated by the squirrels found in the lowland forests of West Malaysia. Of the 25 Malaysian species 11 are nocturnal and the rest active in daytime and the latter can be divided into terrestrial, arboreal and climbing categories with different species showing different use of the various forest strata. The Three-striped ground squirrel and Long-nosed squirrel feed on the ground or eat fallen wood whereas the Slender squirrel is most active on the tree trunks of the lower forest levels. The Plantain and Horse-tailed squirrels travel and feed mainly in the lower and middle forest levels but nest in the upper canopy. The three largest species live highest in the canopy.

The Malaysian squirrels show considerable divergence in choice of food when food is abundant but considerable overlap when it is scarce (all species then rely heavily on bark

and sap). None of the smaller forest squirrels apart from the Horse-tailed squirrel are seed specialists (unlike African or temperate forest species of comparable size). The Three-striped ground squirrel feeds on plant and insect material and the Long-nosed squirrel is an insectivore; these species overlap somewhat with the tree shrews rather than with the arboreal squirrels. The Sunda squirrels (*Sundasciurus* species) feed mainly on bark and sap and most of the beautiful squirrels (genus *Callosciurus*) are opportunistic feeders on a variety of plant material, supplemented in the smaller species with insects. The larger flying squirrels eat a higher proportion of leaves than the smaller species which take mainly fruit.

The three largest species of squirrels active by day show less clear ecological divergence than the smaller species of the lower forest levels. All three are fruit-eaters but the Cream giant squirrel shows more use of the middle canopy levels and takes a significant proportion of leaves in its diet. The Black giant squirrel and Prevost's squirrel are often seen feeding together at the same fruit trees but Prevost's squirrel eats a smaller range of fruit. They also have different foraging patterns. While the larger giant squirrel is at an advantage in competitive situations its travel and basic metabolism are more expensive in energy, whereas Prevost's squirrel can afford to spend less time feeding each day and can travel further to food trees.

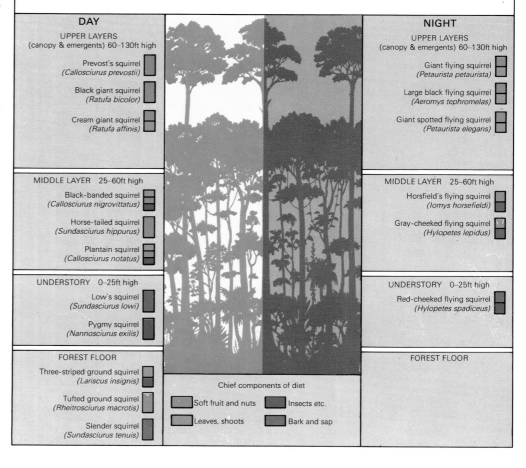

DAY

UPPER LAYERS
(canopy & emergents) 60–130ft high

Prevost's squirrel
(*Callosciurus prevostii*)

Black giant squirrel
(*Ratufa bicolor*)

Cream giant squirrel
(*Ratufa affinis*)

MIDDLE LAYER 25–60ft high

Black-banded squirrel
(*Callosciurus nigrovittatus*)

Horse-tailed squirrel
(*Sundasciurus hippurus*)

Plantain squirrel
(*Callosciurus notatus*)

UNDERSTORY 0–25ft high

Low's squirrel
(*Sundasciurus lowi*)

Pygmy squirrel
(*Nannosciurus exilis*)

FOREST FLOOR

Three-striped ground squirrel
(*Lariscus insignis*)

Tufted ground squirrel
(*Rheithrosciurus macrotis*)

Slender squirrel
(*Sundasciurus tenuis*)

NIGHT

UPPER LAYERS
(canopy & emergents) 60–130ft high

Giant flying squirrel
(*Petaurista petaurista*)

Large black flying squirrel
(*Aeromys tephromelas*)

Giant spotted flying squirrel
(*Petaurista elegans*)

MIDDLE LAYER 25–60ft high

Horsfield's flying squirrel
(*Iomys horsefieldi*)

Gray-cheeked flying squirrel
(*Hylopetes lepidus*)

UNDERSTORY 0–25ft high

Red-cheeked flying squirrel
(*Hylopetes spadiceus*)

FOREST FLOOR

Chief components of diet

Soft fruit and nuts   Insects etc.

Leaves, shoots   Bark and sap

▲ **Appealing but destructive.** The Least chipmunk, like other chipmunks, can be tamed and kept as a pet. Its natural habitat, however, is semiopen land where it can grub up newly planted corn seed in spring and invade granaries in fall.

▶ **Inverted heavyweight.** The Malabar squirrel and the other three species of giant squirrels (all of which live in Southeast Asia) can weigh up to 6.6lb (3kg). Yet for eating this is the characteristic position adopted: hung by the hind feet from a branch.

◀ **Strata use** by squirrels in a Bornean forest (see Box Feature).

areas and lead to the collapse of irrigation channels.

Those species of squirrels that live in colder climates are often hunted for their thick soft winter fur. Marmots play an important role in the Mongolian economy where they are hunted for fur and their meat. In the USSR sousliks, European flying squirrels and Red squirrels are important to the fur trade: the best furs come from the Taiga where the Red squirrels have dark gray winter coats. Gray squirrel tails are used in artists' brushes and marmot fat is used in the Alps as a remedy for chest and lung diseases.

Squirrels can also be carriers of disease. The fleas on the fur of sousliks can carry the plague bacillus and both sousliks and marmots may spread this disease. Golden-mantled ground squirrels are also carriers of bubonic plague and tularemia and marmots may carry Rocky mountain tick fever. For most people, however, squirrels are not considered as a source of disease but as a charming and amusing addition to our woodlands and prairies.   KM

The Role of Kinship

The annual round in Belding's ground squirrels

Belding's ground squirrels (*Spermophilus beldingi*) are meadow-dwelling rodents that inhabit mountainous regions in the western United States. They are active above ground during the day and spend the night in subterranean burrows. While primarily vegetarian, and especially fond of flower heads and seeds (*Spermophilus* means "seed loving"), they also eat insects, birds' eggs, carrion and occasionally Belding young

A population of these animals located at Tioga Pass, high in the central Sierra Nevada of California (9,945ft, 3,040m) has been studied for 15 consecutive years. There ground squirrels are active only from May to October, hibernating the rest of the year. In the spring males emerge first, a few weeks before females; to reach the surface they must often tunnel through several feet of accumulated snow. Once the snow melts, females emerge and the annual cycle of social and reproductive behavior begins.

Females mate about a week after they emerge. During a single afternoon of sexual receptivity each female typically mates with 3–5 different males; studies of paternity have revealed that most litters (55–78 percent) are sired by more than one male. In the presence of receptive females, males threaten, chase and fight with each other; often they sustain physical injuries. The heaviest, oldest, most experienced males usually win such conflicts. They thereby remain near receptive females, enabling them to mate frequently. The majority of males, however, seldom or never copulate.

After mating each female digs a nest burrow in which she rears her litter. Gestation lasts about 24 days, lactation 27 days. The mean litter size is 5, with a range of 1–11. A typical nest burrow is 10–16ft (3–5m) long, 11.5–19.5in (30–50cm) below ground, and has at least two surface openings. Females shoulder the entire parental role. Indeed males often do not even interact with the young of the year because by the time weaned juveniles begin to emerge above ground in late-July or August some males are already reentering hibernation. The females begin to hibernate early in the fall and finally, when it begins to snow, the young of the year emerge to begin their first long, risky winter.

The 7–8 month hibernation period is a time of heavy mortality. Two-thirds of the juveniles and one-third of the adults perish during the winter. Most die because either they deplete their stores of body fat and freeze to death or else they are eaten by predators. Males typically live 2–3 years, compared to 3–4 years for females; a few

males survive 6 years but many females live at least 10 years. Males apparently die younger due to injuries incurred during fights over females and because males that are disabled by infected wounds are particularly susceptible to predators.

Dispersal also differs between the sexes: while females are sedentary from birth, males are relatively nomadic. Soon after they are weaned, juvenile males disperse from the area in which they were born. Typically they move 975–1,300ft (300–400m) and they rarely, if ever, return home or associate again with their maternal relatives. In mid-summer, when females are lactating, many adult males also emigrate. These dispersal episodes virtually preclude social interactions and inbreeding between males and their close kin. Females, in contrast, never disperse and they remain close to their natal burrow and interact with maternal kin throughout their lives.

The ground squirrels' population structure has set the stage for the evolution of nepotism (the favoring of kin). There are four major manifestations of nepotism among females. First, they seldom chase or fight with their close relatives—offspring, mother or sisters—when establishing nest burrows. Among kin, females thus obtain residences with minimum expenditure of energy and little danger of injury. Second, close relatives share portions of their nesting territories and permit each other access to food and burrows on such areas. Third, close kin join together to evict distant relatives or nonkin from each other's territories. Fourth,

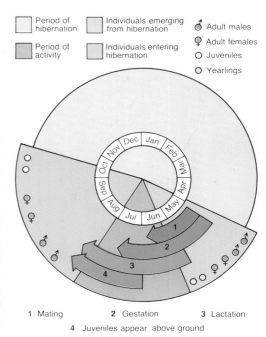

Period of hibernation	Individuals emerging from hibernation	♂ Adult males
Period of activity	Individuals entering hibernation	♀ Adult females
		○ Juveniles
		○ Yearlings

1 Mating 2 Gestation 3 Lactation
4 Juveniles appear above ground

▲ **The annual cycle** of hibernation, activity, and breeding in the Belding's ground squirrel at Tioga Pass, California.

▶ **Alarm call.** A female Belding's ground squirrel calls as a coyote approaches. The function of such calls is apparently to warn offspring and other close relatives of impending danger.

▶ **First day above ground.** BELOW These pups are about 27 days old and have just emerged from their natal burrow.

▼ **Grass for the nest,** carried by a female Belding's ground squirrel. To line one nest can require more than 50 loads of dry grass.

females give warning cries when predatory mammals appear. For example, at the sight of a badger, coyote or weasel some ground squirrels stand erect and give loud, staccato alarm calls. More callers than noncallers are attacked and killed by predators, so calling is self-sacrificial or "altruistic" behavior. However, not all individuals take the same risks. The most frequent callers are old (4–9 years), lactating, resident females with living offspring or sisters nearby. Males and young (1 year) or barren females call infrequently. In other words, callers behave as if they are trading the risks of exposure to predators for the safety and survival of dependent kin.

Cooperative defense of the nest burrow is another important manifestation of nepotism. During gestation and lactation females defend the area surrounding their nest burrow against intrusion by all but their close kin. Such territoriality helps protect the helpless pups from other Belding's ground squirrels: when territories are left unattended, even briefly, unrelated females or one-year-old males may arrive and kill pups. Yearling males typically eat the carcasses, so their infanticide may be motivated by hunger. A different stimulus causes females to become infanticidal. When all of a mother's own young are killed by coyotes or badgers, she often emigrates to a new, safer site. Upon arrival she attempts to kill young there. Only if she is successful is she able to settle. By removing juveniles (females) who are likely to remain in the preferred area, infanticidal females reduce future competition for a nest site. Mothers with close relatives as neighbors lose fewer offspring to infanticide than do females without neighboring kin. This is because groups of females detect marauders more quickly and expel them more rapidly than do individuals acting alone, and because pups are defended by their mother's relatives even when she is temporarily away from home.   PWS

The Squirrel's Devastation

Why do Gray squirrels strip bark?

Many mammals strip living bark from tree trunks. Some, such as deer, rabbits and domestic livestock, eat the tough outer bark, which tends to be rich in mineral nutrients, especially zinc. Other mammals, including some bears, dormice and several species of squirrel, remove and discard the outer bark and then scrape off and eat the soft, sap-containing tissues underneath. These are the tree's circulatory system, which transports water and dissolved minerals up from the roots, and sugars or other organic compounds down from the leaves in the crown of the tree. Although trees tolerate a limited amount of bark-stripping, and may eventually cover wounds with callus, the removal of bark right round the trunk cuts the flow of nutrients to and from the crown. When this happens low down, the whole tree will die, whereas a ring higher up may cause the upper part alone to succumb and the tree will become stunted. A tree not killed outright risks attack by insects or fungi on its exposed heartwood, and even wounds that heal leave inner crevices which spoil the tree's value as timber.

The Gray squirrel (*Sciurus carolinensis*) was introduced to Britain in 1876 from the deciduous woodlands of North America, where it occasionally causes damage by stripping bark from sugar maples. In Britain, Gray squirrels damage mainly sycamore (which is also a maple), beech and oak, although other trees are sometimes attacked. The native Red squirrel (*Sciurus vulgaris*), which has been replaced by the closely related Gray in most deciduous woods, also sometimes strips these species (as well as damaging conifers). Both squirrels strip bark from the stems and large side branches while the trees are still quite young, usually with a diameter less than about 8in (20cm) at the base and not more than 30 or 40 years old. On older trees the bark becomes too thick for easy removal, except on the smaller branches.

Squirrels are sometimes seen peeling fine strips of bark from the outer twigs of limes, thujas and some other trees, but this is taken as a soft lining for tree nests (dreys) or for dens in hollow trees. They do not hibernate during the winter but in Britain in the coldest months they usually forage for only 1–3 hours each day, conserving their body heat at other times in their well-insulated dreys and dens. Squirrels in these nests may also benefit from the shared warmth of several others: up to seven in one drey.

Because squirrel damage is costly and discourages the planting of native beech and oak woods, it is important to find effective ways to prevent bark-stripping. One solution is to kill squirrels in areas with vulnerable young trees. However, shooting and trapping can be expensive, and do not always remove enough squirrels to prevent damage. Poisoning with warfarin is more effective, but may harm other wildlife. Another approach has been to study the causes of the damage, in order to find cheaper solutions to the problem.

There have been many attempts to explain bark-stripping. One was that squirrels might have an uncontrolled gnawing reflex, rather like a dog's tendency to keep scratching with its hind leg after being

▶ **Destroyed in its prime.** Young trees with relatively thin bark, such as this sycamore, succumb most easily to the stripping activity of Gray squirrels.

▼ **Trees that survive,** especially oak, may well house squirrel dreys. Gray squirrels depend on oak trees for supplies of acorns, yet this does not exclude oaks from attack.

tickled behind an ear. Another suggestion was that gnawing might be necessary to prevent the incisor teeth growing too long. As in other rodents, squirrel incisor teeth grow continuously throughout life, and there are no hard nuts to chew in mid-summer, when most damage occurs. These theories can be discounted, because the squirrels do not simply gnaw but peel and eat. The theory that squirrels are marking territory boundaries must also be wrong, because although Gray squirrels share overlapping ranges in which they have a "pecking order" they do not defend specific territories. A fourth idea, that squirrels eat

sap to obtain water, leaves unexplained the fact that damage often occurs next to ponds or streams.

Any theory that attempts to explain bark-stripping must also show why the damage does not occur in all areas of young woodland, nor every year in the areas that are damaged. One possibility is that squirrels are short of food in particular areas or years during mid-summer: tree flowers and buds provide an abundance of food in spring, but food may then be less plentiful for several weeks before autumn nuts and mast are ripe. Another theory is that squirrels simply like the sweet sap.

Strange as it may seem, recent research strongly supports the last suggestion. Individual trees in a wood differ greatly in the thickness of their sappy tissue, even when their age is similar; the squirrels strip most bark from the trees with the thickest sap. Moreover, trees in some woods tend to be more sappy than in others; the squirrels cause most damage in the areas where the trees in general have thick sap. There is little bark-stripping where trees have little sap. Since sap thickness varies from year to year in the same area, this probably explains why the amount of bark-stripping is so variable. It also explains the forester's lament, "that they always ruin the best trees," since it is the trees with most sap, and thus with most circulating nutrients, that are growing most strongly.

Although the extent of damage in an area seems to depend mainly on the quality of its trees, there is also evidence that aggressive behavior may sometimes trigger the bark-stripping. Spring breeding tends to be best in the worst-affected areas, so that gnawing-displays may be most likely there, leading to the discovery of the sweet sap. Gnawing in search of food may help squirrels to discover the sap too, because squirrels tend to have lost most weight in the areas with worst damage. Sap is not a rich food, however, and damage can be quite heavy and yet provide no more than one day's food for five or six squirrels. Nevertheless, whatever triggers the damage, it is probably least likely to start if there are few squirrels in an area.

An improved understanding of the damage indicates new ways of trying to prevent it. Where tree growth does not need to be rapid, in nature reserves for instance, we can try to regenerate beech without damage by making sure that the young trees grow densely, and therefore have a poor sap flow until they are old enough for their bark to resist squirrels. We may eventually be able to breed oak or beech trees with unpleasant-tasting sap, or to tempt squirrels away from them with artificial "sap" dispensers. Moreover, we can probably keep squirrel numbers down by planting vulnerable young trees well away from mature oaks, chestnuts, hazels and conifers, which provide abundant winter food for squirrels. In our existing woodlands, however, there is often no alternative to killing squirrels, so perhaps future research should concentrate on how best to prepare squirrel pie. REK

POCKET GOPHERS

Family: Geomyidae
Thirty-four species in 5 genera divided into two tribes.
Distribution: N and C America, from C and SW Canada through the W and SE USA and Mexico to the Panama–Colombia border.

Habitat: friable soils in desert, scrub, grasslands, montane meadows and arid tropical lowlands.

Size: ranges from head-body length 4.7–8.9in (12–22.5cm) and weight 1.6–14oz (45–400g) in the genus *Thomomys* to head-body length 7–11.8in (18–30cm) and weight 11–32oz (300–900g) in the genus *Orthogeomys*. (Males are always larger than females, up to twice the weight.)

Gestation: 17–20 days in the genera *Thomomys* and *Geomys*.

Longevity: maximum 4 years (6 years recorded in captivity).

Tribe Geomyini

Eastern pocket gophers
Genus *Geomys*, 5 species including: **Plains pocket gopher** (*G. bursarius*).

Taltuzas
Genus *Orthogeomys*, 10 species.

Yellow and cinnamon pocket gophers
Genus *Pappogeomys*, 9 species including: **Yellow-faced pocket gopher** (*P. castanops*).

Michoacan pocket gopher
Zygogeomys trichopus.

Tribe Thomomyini

Western pocket gophers
Genus *Thomomys*, 9 species including **Valley pocket gopher** (*T. bottae*), **Northern pocket gopher** (*T. bulbivorus*).

Pocket gophers present a paradox: common to all species and genera is a basic body plan and a similar life cycle, both necessary for their life of digging, yet one of the most characteristic features of the family is an extensive variability in details of biology, both between species and within them.

In the United States the term "gopher" is used not only for the members of the family Geomyidae but is also applied to some ground squirrels (genus *Spermophilus*), to salamanders and to mud turtles; it suggests a way of life in which digging is important. "Pocket" refers to the external, fur-lined cheek pouches located on either side of the pocket gopher's mouth, a feature of the body which, among mammals, is only otherwise found in the closely related family Heteromyidae (pocket mice and kangaroo rats).

Highly modified for subterranean digging, pocket gophers have thickset, tubular bodies with little external evidence of a neck; short and powerful fore and hind limbs of approximately equal size; and a short, nearly naked tail, which is particularly sensitive to touch. Both upper and lower incisors project through the furred lips so that these teeth are exposed even when the mouth is closed. This adaptation enables pocket gophers to cut roots or dig with their teeth without dirt entering the mouth cavity. They excavate soil with the enlarged claws of their forefeet; the incisors are secondary implements. Populations or species living in harder soils tend to have more forward-pointing incisors and probably make more use of them for digging than do those living in looser soils.

Their skin fits loosely. It is usually clothed in short, thick fur through which are scattered hairs sensitive to touch. The loose skin enables individuals to execute tight turns in the constricted space of their burrows. Gophers are very agile and can move rapidly both backwards and forwards.

The pocket gopher's skull is massive and strongly ridged, with massive jaw muscles. Its incisors may be either smooth or grooved on their anterior surface, according to the genus. The four cheekteeth in each quarter of the jaw form a particularly effective battery for grinding tough and abrasive foodstuffs. All the cheekteeth grow continuously. Indeed the rate of growth of both incisors and foreclaws is amazingly rapid, from 0.02 to 0.04in (0.5–1mm) per day in the Valley pocket gopher.

Within the range of their distribution pocket gophers are ubiquitous in virtually all habitats with extensive patches of friable soil. The range of habitats in which they are found can be extreme: the Valley pocket gopher ranges from desert soils below sea level to alpine meadows well above the timberline, over 11,500ft (3.500m). In tropical latitudes populations of the same species may be found in mountain forest meadows or in arid tropical scrub, but few penetrate true tropical savanna.

Both genera and species are distributed contiguously. Except in very narrow zones of overlap only one kind of pocket gopher will be found in any particular area. This results from an apparent inability to subdivide the two-dimensional subterranean habitat (the fossorial niche). In areas where several species and/or genera meet the pattern of species distribution is mosaic-like. In the mountains of the western USA and Mexico species live in a succession of zones arranged according to altitude.

The two tribes of living pocket gophers represent major branches in the evolutionary divergence of their family. Modern genera first appear in the fossil record in the late Pliocene era (about 4 million years ago). The evolutionary history of the family can be characterized as successive attempts to invade the fossorial niche. Each successive group shows better developed adaptations for subterranean existence. The modern pocket gophers, all members of the subfamily Geomyinae (two other subfamilies are extinct), display the most strongly devel-

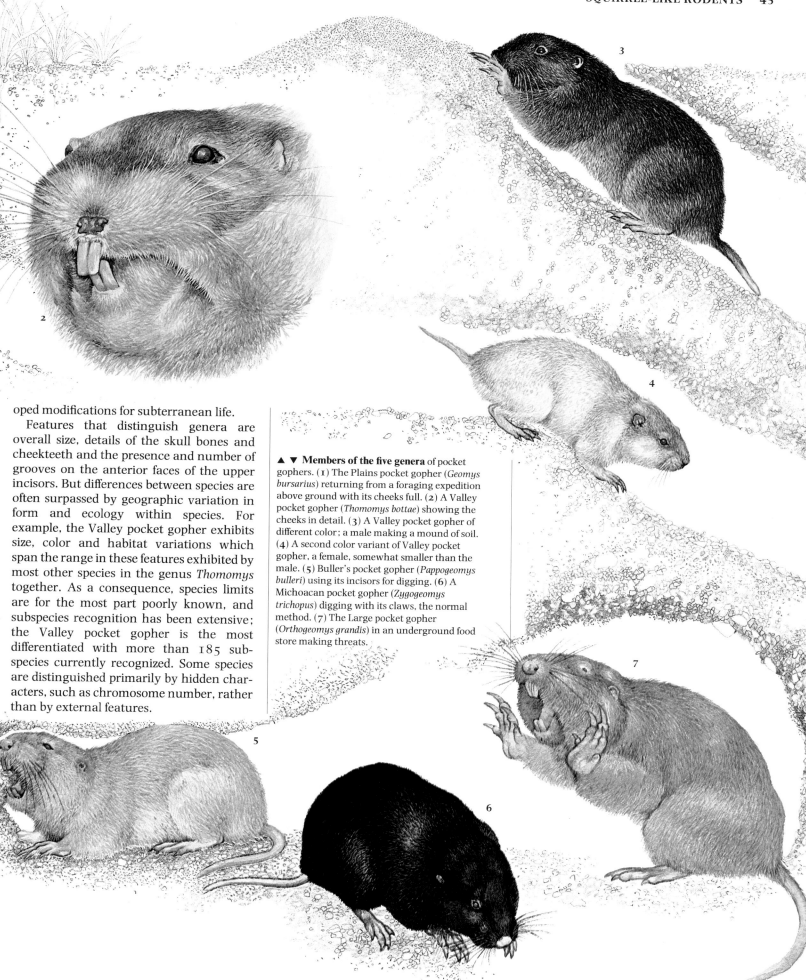

oped modifications for subterranean life.

Features that distinguish genera are overall size, details of the skull bones and cheekteeth and the presence and number of grooves on the anterior faces of the upper incisors. But differences between species are often surpassed by geographic variation in form and ecology within species. For example, the Valley pocket gopher exhibits size, color and habitat variations which span the range in these features exhibited by most other species in the genus *Thomomys* together. As a consequence, species limits are for the most part poorly known, and subspecies recognition has been extensive; the Valley pocket gopher is the most differentiated with more than 185 sub-species currently recognized. Some species are distinguished primarily by hidden char-acters, such as chromosome number, rather than by external features.

▲ ▼ **Members of the five genera** of pocket gophers. (1) The Plains pocket gopher (*Geomys bursarius*) returning from a foraging expedition above ground with its cheeks full. (2) A Valley pocket gopher (*Thomomys bottae*) showing the cheeks in detail. (3) A Valley pocket gopher of different color; a male making a mound of soil. (4) A second color variant of Valley pocket gopher, a female, somewhat smaller than the male. (5) Buller's pocket gopher (*Pappogeomys bulleri*) using its incisors for digging. (6) A Michoacan pocket gopher (*Zygogeomys trichopus*) digging with its claws, the normal method. (7) The Large pocket gopher (*Orthogeomys grandis*) in an underground food store making threats.

Pocket gophers eat a wide variety of plants. Above ground they take leafy vegetation from the vicinity of burrow openings; underground they devour succulent roots and tubers. They often prefer forbs and grasses, but diet shifts seasonally according to the availability of food and the gophers' needs for nutrition and water. For example, water-laden cactus plants may become a major dietary component during the hot and dry summer months in arid habitats. The external cheek pouches are filled with parts of plants by dextrous motions of the forepaws for transportation to cache storage areas in the burrow. The storage areas are usually sealed from the main tunnel system.

Males do not become sexually mature until the breeding season after their birth, ie when about one year old. Females in some populations, however, may start to breed during the season of their birth, when approximately 70 days old. For most species and populations the reproductive period is bounded by the changes of the seasons, but in its timing it varies geographically. In montane regions breeding follows the melting snow, but in coastal and desert valleys and in temperate grasslands it coincides with winter rainfall.

The number of litters per female may vary geographically within species. In the Valley pocket gopher many populations have but one litter per year while in others a given female may have several successive litters. Pocket gophers of this species, living in irrigated agricultural fields, may breed year-round though neighboring populations living amidst natural vegetation may have sharply delimited seasonal breeding. Females of Yellow-faced and Plains pocket gophers typically produce multiple litters each year. Litter size varies from *Pappogeomys* in which two is the commonest number of young to *Thomomys* which usually produces five. Some females, however, bear up to 10 young per pregnancy. The onset of breeding, length of season and number of litters are largely controlled by local environmental conditions, primarily temperature, moisture and vegetation.

Young are born completely dependent on their parents. Their cheek pouches do not open for about 24 days; eyes and ears open at 26 days. In Northern and Valley pocket gophers the molting from juvenile coats takes 100 days. Weaning in most gophers probably occurs by 40 days after birth, but dispersal from the mother's burrow system may not occur for about 60 days. The maximum longevity in nature is about 5 years, but the average life span of an adult is considerably shorter, just over one year. Females live nearly twice as long as males. On average females survive for 56 weeks and males for 31 weeks in the Yellow-faced pocket gopher while in the Valley pocket gopher longevity is 4.5 years for females and 2.5 years for males. The ratio between adult numbers of the sexes varies geographically from near equality to a high preponderance of females (3 to 4 females for every male) in probably all species, although this observation is best documented for Valley, Northern, Plains and Yellow-faced pocket gophers. Such variation is related to geographic differences in population density, with males becoming proportionately less common as the population size increases. In part the proportional loss of adult males is thought to result from increased mortality due to male-male aggressive encounters over territorial space and mates. Virtually all adult females within a population will breed each season, but there may be high variance in male reproductive success. It is possible that some males never breed during their life span, and that most young in a given population are sired by relatively few.

Pocket gophers are solitary creatures, with individuals regardless of sex maintaining separate and contiguous territories. In the Valley pocket gopher the maximum territory size is about 2,700sq ft (250sq m). The size of territories varies in part according to the quality of habitat: high-quality habitat will support denser populations with smaller individual territory sizes than will habitats of poorer quality. Territorial boundaries break down momentarily during the

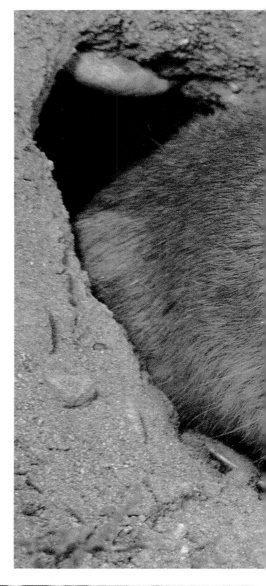

▲ Out of the raw earth, a Northern pocket gopher. Gophers prefer deep, sandy, easily crumbled, well-drained soils, but this species can live in a wider range of soils than any other. Gopher burrows are commonly 50ft (about 15m) long.

◄ The digging machine. This Valley pocket gopher displays well the contrast between gophers' small eyes and ears and their obtrusive claws.

tend to be evenly dispersed, occupying all available habitat. In poor-quality regions density is depressed, individuals tend to be clumped in distribution and their territories will shift in both size and position throughout the year as animals search for food.

Male and female young disperse from the mother's system at the same time, but in the Valley pocket gopher female young of the year initially move longer distances, establishing territories in which they may breed during the season of their birth. Male young tend to live in shallow systems in marginal and peripheral habitat until they disperse to establish territories just prior to the next breeding season. Despite their adaptations for digging, pocket gophers can and do move considerable distances and most dispersal happens above ground, on dark nights. Movements over shorter distances may take advantage of long tunnels just under the surface.

Individuals of both sexes are pugnacious and aggressive, and will fight if placed together in limited quarters. Individuals of both sexes in all species compete for parcels of land, which must contain all their requirements for long-term survival.

Pocket gophers serve a major role in soil dynamics. Their constant digging generates a vertical cycling of the soil, counteracting the packing effects of grazers, making the soil more porous and hence slowing the run-off of water, and providing increased aeration for plant growth. They can have profound effects on plant communities, often, through continual disturbance, creating soil conditions that favor the growth of herbaceous plants which are often preferred foods. The range and population density of many pocket gopher species have increased as native plant communities have been replaced by agricultural development. However, although pocket gophers can benefit agricultural interests by working the soil, they can cause extensive economic loss through their voracious herbivorous appetite and their digging activities. Dense populations can produce severe loss of crops and, particularly in the arid western region of North America, irrigation systems can rapidly be undermined by their extensive burrow systems. Millions of dollars have been spent on programs to control pocket gopher populations in the USA. However, most control efforts have been unable to deal effectively with the increase in population density and reproductive rate in pocket gopher populations which results from their invasion of the high-quality resources provided by agricultural habitat.   JLP

breeding season, at which time there occurs multiple occupancy of burrows by adult females and adult pairs or by females with young. Male territory size in the Valley pocket gopher averages more than twice that of females in both area and total burrow length. Each adult male territory may be contiguous with two or more female burrow systems. While each individual maintains a tunnel system for exclusive use, adjacent males and females may have common burrows and deep, common nesting chambers.

Densities in smaller species such as the Valley pocket gopher will generally not exceed 99 adults per acre (40 per ha) and will be as low as 17 per acre (7 per ha) in the large-bodied Yellow-faced pocket gopher. In high-quality habitat individual territories are stable in both size and position, with most individuals living their entire adult lives in very limited areas. In such conditions densities are high and individuals

SCALY-TAILED SQUIRRELS AND POCKET MICE

Family: Anomaluridae
Scaly-tailed squirrels
Seven species in 3 genera.
Distribution: W and C Africa.

Family: Heteromyidae
Pocket mice
Sixty-five species in 5 genera.
Distribution: N, C and northern S America.

Scaly-tailed squirrels   Pocket mice

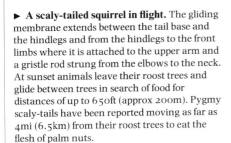

▶ **A scaly-tailed squirrel in flight.** The gliding membrane extends between the tail base and the hindlegs and from the hindlegs to the front limbs where it is attached to the upper arm and a gristle rod strung from the elbows to the neck. At sunset animals leave their roost trees and glide between trees in search of food for distances of up to 650ft (approx 200m). Pygmy scaly-tails have been reported moving as far as 4mi (6.5km) from their roost trees to eat the flesh of palm nuts.

THE tropical and subtropical forests of the Old World are inhabited by an interesting array of gliding mammals. In tropical Africa this niche is filled by members of the Anomaluridae, the gliding **scaly-tailed squirrels,** which apparently share only distant evolutionary relationships with true squirrels.

The scaly-tailed squirrels are squirrel-like in form with a relatively thin, short-furred tail whose underside contains an area of rough, overlapping scales near the base of the tail.

The ecology and behavior of scaly-tailed squirrels is poorly known although such species as Lord Derby's scaly-tailed flying squirrel and Zenker's flying squirrel are common and sometimes come into contact with humans. All species are probably nocturnal, and some spend their days in hollow trees where colonies of up to 100 Pygmy scaly-tailed squirrels have been found. Beecroft's anomalure apparently does not sleep in hollow trees but instead rests on the outside of tree trunks during the day where it relies on its cryptically colored fur for protection from predators. Lord Derby's scaly-tailed flying squirrel, the largest member of the family, eats a variety of plant products, including bark, fruits, leaves, flowers and green nuts, as well as insects. If anomalures are like other groups of rodents, smaller species eat a higher proportion of protein-rich insects than large members.

The scant information available about anomalure reproduction indicates that females may have 2 litters of 1–3 young per year. At birth babies are large, well-furred, active, and their eyes are open. Female pygmy scaly-tails apparently leave their colonies to bear their single young alone.

Except for Derby's anomalure, these rodents depend entirely on primary tropical forest for their existence. To the extent that African primary forests are being destroyed, these interesting but poorly-known rodents are endangered.

By day, summer conditions in the deserts in the southwestern USA are formidable. Surface temperatures soar to over 122°F (50°C), sparsely distributed plants are parched and dry and signs of mammalian life are minimal. As the sun sets, however, the sandy or gravelly desert floor comes alive with a high density of rodents (25–49 animals per acre, 10–20 per ha). Greatest diversity occurs among the **pocket mice** in which five or six species can coexist in the same barren habitat.

Contrast these hot, dry (or in winter, cold and apparently lifeless) desert conditions with the habitat in which species of the genus *Heteromys* can attain densities of up to 44 per acre (18 per ha): the tropical rain forests of Central and northern South America. Rich in rain and in vegetation, the rain forest is yet near bare in heteromyid species: only one species, Desmarest's spiny pocket mouse, occurs, at most sites.

The most likely explanation for this difference in species richness lies in the diversity and availability of seeds. In North American deserts, seeds of annual species can accumulate in the soil (to a depth of 0.8in, 2cm) in densities of up to 8,450 seeds per sq ft (91,000 per sq m). Patchily distributed by wind and water currents, small seeds, weighing about a milligram tend to accumulate in great numbers under shrubs and bushes and on the leeward sides of rocks whereas larger seeds occur in clumps in open areas between vegetation. This soil-seed matrix is the arena for the seed-gathering activities not only of nocturnal heteromyid rodents but also of a rather diverse array of seed-harvesting ants (about 50 species) and a modest diversity of seed-eating birds (four species, active by day).

Tropical forests are also rich in seeds, but many of those produced by tropical shrubs and trees are protected, chemically, against predation by seed-eating insects and their larvae, and by birds and mammals. This is especially true of large seeds weighing several ounces, which considerably reduces the variety of seeds available to rodents. In effect, then, the tropical rain forest is a desert in the eyes of a seed-eating rodent whereas the actual desert is a "jungle" with regard to seed diversity and availability. Most of the 15–20 species of rodents inhabiting New World tropical forests are omnivorous or are fruit-eaters.

With large cheek pouches and a keen sense of smell, heteromyid rodents are admirably adapted for gathering seeds. Most of the time they are active outside their underground burrow systems is spent collecting

▲ **The benefits of hopping.** Kangaroo rats can range over wider areas than pocket mice and therefore can be more selective when foraging. They usually dig and glean seeds from one or two spots under a shrub and then rapidly hop to another shrub and dig again. Foraging away from cover exposes kangaroo rats, such as this Pacific kangaroo rat, to greater risk from such predators as owls, but their hearing—sensitive to low frequencies—enables kangaroo rats to detect predators more readily than other heteromyid rodents can.

Quadrupedal pocket mice are on the other hand almost "filter-feeders": they push slowly through loose soil, usually beneath bushes or shrubs, sorting and stuffing seeds into their cheek pouches as they encounter them. Not all seeds collected will be eaten or stored. For consumption or storage, pocket mice tend to select only seeds rich in calories.

seeds in various locations within their home ranges. Members of the two tropical genera (*Liomys* and *Heteromys*) search through the soil litter for seeds with which they fill their cheek pouches. Some of these will be buried in shallow pits scattered around the home range whereas others will be stored underground in special burrow chambers.

Breeding in desert heteromyids is strongly influenced by the flowering activities of winter plants which germinate only after at least 1in (2.5cm) of rain has fallen between late September and mid December. In dry years, seeds of these plants fail to germinate, and a new crop of seeds and leaves is not produced by the following April and May. In the face of a reduced food supply heteromyids do not breed, and their populations decline in size. In years following good winter rains most females produce two or more litters of up to five young, and populations increase rapidly in size.

This "boom or bust" pattern of resource availability also influences heteromyid social structure and levels of competition between species. When seed availability is low, seeds stored in burrow or surface caches become valuable and defended resources. Behavior becomes asocial in most species of arid-land heteromyids (including species of *Liomys*): adults occupy separate burrow systems (except for mothers and their young) and when two members of a species meet away from their burrows they engage in "boxing matches" and sand kicking fights. In the forest, in contrast, species of *Heteromys* are socially more tolerant; individuals have widely overlapping home ranges, they share burrow systems and are less likely to fight each other.

Experiments conducted in Arizona, USA, indicate that heteromyids not only compete for seeds among themselves but also with seed-harvesting ants. In one set of experiments heteromyid numbers and biomass increased 20–29 percent above levels on control plots after ant colonies had been removed from plots of 0.25 acre (0.1ha). Similarly, the total number of ant colonies increased 71 percent over control levels on plots from which heteromyids had been removed. Exclusion of large kangaroo rats from plots surrounded by "semipermeable" fences which permitted the immigration of small pocket mice resulted in a 3.5-fold increase in pocket mouse numbers after a period of eight months. Densities of small omnivorous rodents (eg *Peromyscus*, *Onychomys*) did not differ between experimental and control plots, which indicates that kangaroo rats influence only the abundance of pocket mice, not of all species of small rodents.

The diversity and availability of edible seeds is the key to the evolutionary success of heteromyid rodents. Seed availability affects foraging patterns, population dynamics and social behavior. In North American deserts, because seed production influences levels of competition among heteromyids, ants and probably other seed-eaters, there is a clear link between resources and the structure of an animal community. Thus the abundance and diversity of seed-eaters is directly related to plant productivity. TEF

The 3 genera of scaly-tailed squirrels

Habitat: tropical and subtropical forests. Size: ranges from head-body length 2.7–3.1in (6.8–7.9cm), tail length 3.6–4.8in (9.1–11.7cm), weight 0.5–0.6oz (14–17.5g) in the Pygmy scaly-tailed squirrel (*Idiurus zenkeri*) to head-body length 10.6–15in (27–37.9cm), tail length 8.6–11.2in (22–28.4cm), weight 16–38oz (450–1,090g) in Lord Derby's scaly-tailed flying squirrel (*Anomalurus derbianus*). Gestation: unknown. Longevity: unknown (probably several years in Lord Derby's scaly-tailed flying squirrel).

Scaly-tailed flying squirrels
Genus *Anomalurus*.
Africa from Sierra Leone E to Uganda and S to N Zimbabwe. Four species including: **Beecroft's anomalure** (*A. beecrofti*), **Lord Derby's scaly-tailed flying squirrel** (*A. derbianus*).

Pygmy scaly-tailed squirrels
Genus *Idiurus*.
Africa from Cameroun SE to Lake Kivu in E Zaire. Two species including **Zenker's flying squirrel** (*I. zenkeri*).

Flightless scaly-tailed squirrel
Zenkerella insignis.
Cameroun.

The 5 genera of pocket mice

Habitat: semiarid to arid regions of N America for species in the genera *Perognathus*, *Microdipodops* and *Dipodomys*; tropical forests and grasslands for *Liomys* and *Heteromys*. Size: ranges from head-body length 2.2–2.4in (5.6–6cm), tail length 1.7–2.3in (4.4–5.9cm) and weight 0.3–0.5oz (10–15g) in *Perognathus flavus* to head-body length 4.9–6.4in (12.5–16.2cm), tail length 7.1–8.5in (18–21.5cm) and weight 3.0–4.8oz (83–138g) in *Dipodomys deserti*. Gestation: ranges from 25 days in *Liomys pictus* to 33 days in *Dipodomys nitratoides*. Longevity: most live only a few months, but up to 5 years have been recorded in the hibernating pocket mouse *Perognathus formosus*.

Kangaroo rats
Genus *Dipodomys*.
SW Canada and USA west of Missouri River to south C Mexico. Bipedal (hind legs long, front legs reduced). Twenty-two species.

Spiny pocket mice
Genus *Heteromys*.
Mexico, C America, northern S America. Quadrupedal. Eleven species including: **Desmarest's spiny pocket mouse** (*H. desmarestianus*).

Spiny pocket mice
Genus *Liomys*.
Mexico and C America S to C Panama. Quadrupal. Five species.

Kangaroo mice
Genus *Microdipodops*.
USA in S Oregon, Nevada, parts of California and Utah. Bipedal (hind legs long, front legs reduced). Two species.

Pocket mice
Genus *Perognathus*.
SE Canada, W USA south to C Mexico. Quadrupedal. Twenty-five species.

SPRINGHARE

Pedetes capensis
Springhare or springhaas.
Sole member of family Pedetidae.
Distribution: Kenya, Tanzania, Angola,
Zimbabwe, Botswana, Namibia, South Africa.

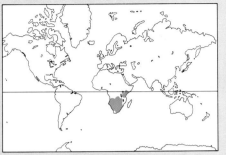

Habitat: flood plains, fossil lake beds, savanna,
other sparsely vegetated arid and semiarid
habitats on or near sandy soils.

Size: head-body length
14–17in (36–43cm); tail
length 16–19in (40–48cm);
ear length 3in (7cm); hind
foot length 6in (15cm);
weight 6–9lb (3–4kg).
(Dimensions are similar for
both female and male.)

Coat: upper parts, lower half of ears, basal half
of tail yellow-brown, cinnamon or rufous
brown; upper half of ears, distal half of tail and
whiskers black; underparts and insides of legs
vary from white to light orange.

Gestation: about 77 days.

Longevity: unknown in wild; more than 14.5
years in captivity.

▼ **Springhare postures:** (1) standing,
(2) foraging on all fours. (3) leaping, (4) grooming.

SCATTERED through the arid lands of East and South Africa are numerous grass-covered flood plains and fossil lake beds. After the rainy season grass is superabundant but eventually, however, it is eaten away by large herbivores and for much of the year it is too short and too sparse for large grazers to forage efficiently. The result is an unused food supply or, to use the zoologist's concept, an empty niche. To fill it requires an animal small enough to use the grass efficiently yet large and mobile enought to travel to the grass from areas that can provide shelter from the weather and from predators. These are attributes of the springhare.

However the springhare still faces formidable problems. It is small enough to be killed by snakes, owls and mongooses, and large enough to be attractive to the largest predators, including lions and man. It seems sensible therefore, to interpret many of the animal's specialized physical and behavioral features as adaptations for an arid environment and for the efficient detection and avoidance of predators.

There is little evidence of the springhare's origins. Some people believe that its closest living relatives are the scaly-tailed flying squirrels (Anomaluridae). The springhare actually resembles a miniature kangaroo. Its hindlegs are very long and each foot pad has four toes, each equipped with a hoof-like nail. Its most frequent and rapid type of movement is hopping on both feet. Its tail is slightly longer than its body and helps to maintain balance while hopping. The front legs are only about a quarter of the length of the hindlegs. Its head is rabbit-like, with large ears and eyes and a protruding nose;

sight, hearing and smell are well developed. When pursued by a predator a springhare can leap 10–13ft (3–4m). When captured it attempts to bite the predator with its large incisors and to rake it with the sharp nails of its powerful hind feet.

Though springhares are herbivorous, they occasionally eat mineral-rich soils and accidentally ingest insects. They are very selective grazers, preferring green grasses high in protein and water.

Springhares are active above ground at all times of night. Normally they forage within 820ft (250m) of their burrows but occasionally they travel as far as 1,300ft (400m). While foraging they are highly vulnerable to predators because they are completely exposed to detection and are far from the safety of their burrows. On nights with a full moon they appear to be particularly vulnerable and move only an average of 13ft (4m) onto the feeding area: by contrast on moonless nights they move on average 190ft (58m). When above ground springhares spend about 40 percent of their time in groups of two to six animals, pre-

sumably because a group is more efficient than an individual at detecting predators.

Springhares spend the hours of daylight in burrows located in well-drained, sandy soils. Burrows lie about 31in (80cm) deep, have 2–11 entrances and vary in length from 33 to 151ft (10–46m). Each burrow is occupied by one springhare or by a mother and an infant. Burrows provide considerable protection against the arid environment and against predators. Some predators, eg snakes and mongooses, can enter burrows however, so springhares often block entrances and passageways with soil after entering. The tunnels and openings in the burrow system provide many escape routes when predators do enter. The absence of chambers and nests within the burrow suggests that springhares do not rest consistently in any one location within the burrow—probably another precaution against predators.

In springhare populations the number of males equals the number of females. There is no breeding season and about 76 percent of the adult females may be pregnant at any one time. Adult females undertake about three pregnancies per year, each resulting in the birth of a single, large well-developed infant.

New-born are well furred and able to see and move about almost immediately, yet they are confined to the burrow and are completely dependent on milk until half grown when they can become completely active above ground. Immature springhares usually account for about 28 percent of all individuals active above ground.

Although the reproductive rate of springhares is surprisingly low, there are two distinctive advantages to be found in the springhare's reproductive strategy. First, the time and energy alloted to the female springhare for reproduction is funneled into a single infant. This results apparently in low infant and juvenile mortality. When the juvenile springhare first emerges from its burrow its feet are 97 percent and its ears 93 percent of their adult size: it is almost as capable of coping with predators and other environmental hazards as a fully grown adult. Second, in having to provide care and nutrition for only one infant the mother is subject to minimal strain. Females that can remain in good physical condition and avoid predators and disease during breeding are most likely to survive to breed again.

Springhares are generally common where they occur, even when they are frequently hunted by man. In the best habitats there may be more than 4 springhares per acre (10 per ha). However, when arid, ecologically sensitive areas are overgrazed by domestic stock, as occurs in the Kalahari Desert, springhare densities decrease in response to decrease in the supply of food.

Springhares are of considerable importance to man as a source of food and skins. In Botswana springhares are the most prominent wild animal in the human diet. A single band of bushmen may kill more than 200 springhares in one year. However, the springhare can also be a significant pest to agriculture, feeding on a wide variety of crops including corn, peanuts, sweet potatoes and wheat.   TMB

◄ **Caught.** An African bushman displays his spoil. No part of a dead Springhare goes to waste. Over 60 percent is eaten, including the eyes, brain and contents of the intestine. The skin is used to make bags, clothing and mats, the long sinew from the tail is used as thread; the fecal pellets are smoked.

▼ **As if a miniature kangaroo,** a Springhare sits upright, showing its short front limbs and powerful hindlegs.

MOUSE-LIKE RODENTS

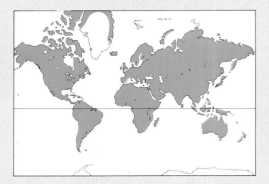

Suborder: Myomorpha
Five families: 264 genera: 1,137 species
Distribution: worldwide except for Antarctica.

Habitat: all terrestrial habitats except snow-covered mountain peaks and extreme high arctic.

Size: head-body length from 1.6in (4cm) in dwarf jerboas to 19in (48cm) in Cuming's slender-tailed cloud rat; weight from 0.2oz (6g) in the Pygmy mouse to 4.4lb (2kg) in Cuming's slender-tailed cloud rat.

Rats and mice
Family: Muridae
One thousand and eighty-two species in 241 genera and 15 subfamilies.

New World rats and mice
Subfamily: Hesperomyinae
Three hundred and sixty-six species in 69 genera.

Voles and lemmings
Subfamily: Microtinae
One hundred and ten species in 18 genera.

Old World rats and mice
Subfamily: Murinae
Four hundred and eight species in 89 genera.

Blind mole-rats
Subfamily: Spalacinae
Eight species in 2 genera.

African pouched rats
Subfamily: Cricetomyinae
Five species in 3 genera

African swamp rats
Subfamily: Otomyinae
Thirteen species in 2 genera.

Crested rat
Subfamily: Lophiomyinae
One species.

African climbing mice
Subfamily: Dendromurinae
Twenty-one species in 10 genera.

Root or bamboo rats
Subfamily: Rhizomyinae
Six species in 3 genera.

Madagascan rats
Subfamily: Nesomyinae
Eleven species in 8 genera.

Oriental dormice
Subfamily: Platacanthomyinae
Two species in 2 genera.

MORE than a quarter of all species of mammals belong to the suborder of Mouse-like rodents (Myomorpha). They are very diverse—difficult to describe in terms of a typical member. However, the Norway rat and the House mouse are fairly representative of a very large proportion of the group, both in overall appearance and in the range of size encountered. Like these familiar examples the great majority of mouse-like rodents are small, terrestrial, prolific, nocturnal seedeaters. The justification for believing them to comprise a natural group, derived from a single ancestor separate from the other suborders of rodents, is debatable but lies mainly in two features: the structure of the chewing muscles of the jaw and the structure of the molar teeth.

The way in which the lateral masseter muscle, one of the principal muscles that close the mouth, passes from the lower jaw up through the orbit and from there forwards into the muzzle is not only unique among rodents but is not found in any other mammals. The structure of the teeth is more variable but there is never more than one premolar tooth in front of the molars.

The great majority of mouse-like rodents belong to the family Muridae, the mouse family. The minority groups are the dormice (family Gliridae) and the jerboas and jumping mice (families Dipodidae and Zapodidae). These represent early offshoots that have remained limited in number of species and also somewhat specialized, the dormice being arboreal and (in temperate regions) hibernating, the jerboas being adapted for the desert. The members of the mouse family (murids) have undergone more recent and

much more extensive changes (adaptive radiation) beginning in the Miocene period, ie within the last 20 million years. Some of the resultant groups are specialists, for example the voles and lemmings which are adapted to feeding on grass and other tough but abundant vegetation. However, many have remained rather versatile generalists, feeding on seeds, buds and sometimes insects, all more nutritious but less abundant than grass.

The greatest proliferation and diversification of species that has ever taken place in the evolution of mammals has occurred in the mouse family, which has over 1,000 living species. Its members are found throughout the world, in almost every terrestrial habitat. They are often the dominant small mammals in these habitats. Those that most closely resemble the common ancestor of the group are probably the common mice and rats found in forest habitats world-wide, typified by the European Wood mouse and the very similar, although not very closely related, American Deer mouse. These are versatile animals, predominantly seedeaters but capable of using their seedeating teeth to exploit many other foods, such as buds and insects.

From such an ancestor many more specialized groups have arisen, capable of exploiting more difficult habitats and food sources. Most gerbils (subfamily Gerbillinae) have remained seedeaters but have adapted to hot arid conditions in Africa and Central Asia. The hamsters (subfamily Cricetinae) have adapted to colder arid conditions by perfecting the arts of food storage and hibernation; the voles and lemmings (sub-

► **The telling face of a rare rat.** The False water rat, a species belonging to the rat and mouse subfamily of Australian water rats and their allies, is seldom seen. A nocturnal creature, it lives on the edge of coastal swamps in northern Australia where it climbs trees rather than swims. Yet for all its obscurity its pointed face, bead eyes and bristling whiskers are instantly recognizable as those of a rat.

▼ **Distinguishing feature** of mouse-like rodents. Both lateral (green) and deep (blue) masseter muscles are thrust forward, providing very effective gnawing action, the deep masseter passing from the lower jaw through the orbit (eye socket) to the muzzle. Shown here is the skull of the muskrat.

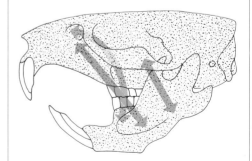

▷ **A more familiar mouse-like rodent.**
OVERLEAF The Wood mouse, also known as the Common field or Long-tailed field mouse, can be found across Europe and Central Asia in a wide variety of habitats—woodlands, fields, hedgerows, gardens, even in buildings. Success is the result of versatility. The Wood mouse is extremely energetic, agile and omnivorous. If necessary it can swim.

Zokors or Central Asiatic mole-rats
Subfamily: Myospalacinae
Six species in 1 genus.

Australian water rats
Subfamily: Hydromyinae
Twenty species in 13 genera.

Hamsters
Subfamily: Cricetinae
Twenty-four species in 5 genera.

Gerbils
Subfamily: Gerbillinae
Eighty-one species in 15 genera.

Dormice
Families: Gliridae, Seleviniidae
Eleven species in 8 genera.

Jumping mice and birchmice
Family: Zapodidae
Fourteen species in 4 genera.

Jerboas
Family: Dipodidae
Thirty-one species in 11 genera.

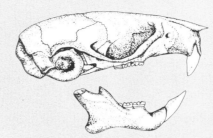

Harvest mouse   0.6in

European hamster   **Libyan jird**

Harvest mouse   **Norway lemming**

Skull and teeth of mouse-like rodents.

Most mouse-like rodents have only three
cheekteeth in each row, but they vary greatly
in both their capacity for growth and the
complexity of the wearing surfaces. The most
primitive condition is probably that found in
the hamsters—low-crowned, with cusps
arranged in two longitudinal rows and no
growth after their initial eruption. The rats
and mice of the subfamily Murinae, typified by
the Harvest mouse, are similar but have three
rows of cusps. The gerbils have high-crowned but
mostly rooted teeth, in which the original
pattern of cusps is soon transformed by wear
into a series of transverse ridges. The voles
and lemmings take this adaptation to a tough,
abrasive diet—in this case mainly grass—even
further by having teeth that continue to grow
and develop roots only late in life or, more
commonly, not at all.

family Microtinae) and the superficially similar African swamp rats (subfamily Otomyinae) have cracked the problem of feeding on grass and similar tough herbage and have thereby opened up new possibilities for expansion by feeding on the vegetation that is dominant over very large areas.

Three independent groups, the blind mole-rats (subfamily Spalacinae), the zokors (subfamily Myospalacinae) and the root rats (subfamily Rhizomyinae) have taken to an underground life, emulating the more distantly related pocket gophers in America (family Geomyidae) and the African mole-rats (family Bathyergidae).

A different kind of diversification, and on a smaller scale, is shown by the rats and mice that succeeded in colonizing the island of Madagascar. There they found a whole variety of vacant habitats waiting to be occupied and therefore the Madagascan rodents (subfamily Nesomyinae), although probably derived from a single colonization, have diversified within that one island in much the same way as have their relatives that stayed behind on the African mainland, with a variety of vole-like, mouse-like and rat-like species.

The indigenous American members of the family (the New World rats and mice, subfamily Hesperomyinae), although very numerous as species (about 366), do not show quite the same degree of diversification as the Old World members. In particular there is only one underground species, the Brazilian shrew-mouse, and it is far less adapted to a mole-like existence than the various Old World mole-rats. This role in the Americas is played by other groups, the pocket gophers in North America and the tuco-tucos in South America.

If one looks at any one area, for example Europe, these specialized groups—mice, voles, hamsters, mole-rats—seem so different that it is tempting to treat them as separate families. This is often done, with considerable justification. However, it is more difficult when the problem is considered worldwide: the whole question of the interrelationships between these groups is still very uncertain.

The ancestral member of the mouse family probably had molar teeth similar to those found in many New World mice and hamsters. These have moderately low-crowned teeth with rounded cusps on the biting surface arranged in two longitudinal rows. The typical Old World mice and rats (subfamily Murinae) appear to have evolved from such an ancestor by developing a more complex arrangement of cusps, forming three rows, while retaining most of the other primitive characters. These two groups are often treated as separate families, the Cricetidae (so-called cricetine rodents) and Muridae (so-called murine rodents) respectively. On this basis several other groups like the gerbils and pouched rats are considered as derivations of, or are included in, the Cricetidae. Others, like the voles, African swamp rats and the various underground groups, have their molar teeth so modified for a rough diet, with high crowns and complex shearing ridges of hard enamel, that their origin is less certain, although most are generally believed to be derived from the cricetid rather than the murid pattern. The African climbing mice (subfamily Dendromurinae) have retained low-crowned, cusped teeth, with additional cusps as in the Murinae but forming a different pattern, suggesting an independent origin from the primitive "cricetid" plan.

These various dental adaptations are reflected in other ways also. By comparison with typical forest mice, species living in dense grassland tend to have shorter tails, legs, feet and ears, as in the voles and hamsters. However, those that live on even more open ground, such as the gerbils in dry steppes and semidesert, tend to retain the "mouse" characters of long hind feet and long tail, adapted to the very fast movement required to escape from predators in the absence of cover. They also retain large ears, and many have enormously enlarged bony *bullae* surrounding the inner ear, acting as resonators and enabling them to detect low frequency sounds carried over great distances and so recognize danger and to communicate over long distances in desert conditions where refuges may be far apart.

Some voles, for example those in the genera *Pitymys* and *Ellobius*, make extensive underground tunnels and in these species the extremities are reduced further and the eyes are very small. These features are also found in the tropical root rats (subfamily Rhizomyinae) and are taken to greater extremes in the more completely underground groups—the zokors (subfamily Myospalacinae) and the blind mole-rats (subfamily Spalacinae).

In ecological terms, most mouse-like rodents would be classified as "r-strategists," ie they are adapted for early and prolific reproduction rather than for long individual life spans ("k-strategists"). Although this applies in some degree to most rodents, the rats and mice show it more strongly and generally than, for example, their nearest relatives, the dormice and the jerboas.   GBC

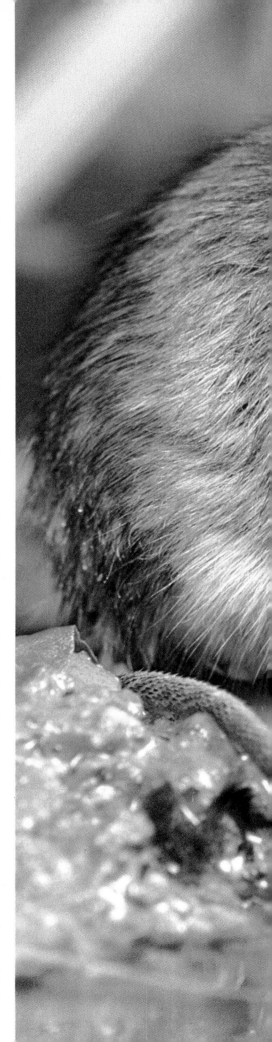

NEW WORLD RATS AND MICE

Subfamily: Hesperomyinae
Three hundred and sixty-six species in 69
genera combined into 13 tribes in 5 groups.
Family: Muridae.
Distribution: N and S America and adjacent
offshore islands.

Habitat: all terrestrial habitats (including
northern forests, tropical forest and savanna)
excluding snow-covered mountain peaks and
extreme high Arctic.

Size: ranges from head-body length 2–3.2in
(5–8.1cm), tail length 1.4–2.2in (3.5–5.5cm),
weight 0.25oz (7g) in the Pygmy mouse to
head-body length 6.3–11.3in (16–28.7cm), tail
length 3–6.3in (7.6–16cm), weight 1lb 11oz
(700g) in the South American giant water rat.

Gestation: 20–50 days.
Longevity: maximum in wild 1 year; some live
up to 6 years in captivity.

▼ **A craving for sticks** is one of the most
distinctive features of the North American
woodrats. Most species, including *Neotoma
micropus* seen here, collect sticks and with them
build piles by their nests. A few desert living
species, however, use cactus instead.

THE New World rats and mice are an
array of 366 species divisible into 69
genera, at present distributed from the
northern forests of Canada south through
central Canada, the USA, Central and South
America to the tip of the continent. They
originated in North America where they
descended from the same kind of primitive
rodents as did the hamsters of Europe and
Asia and the pouched rats of Africa (their
nearest Old World relatives today). These
ancestors, cricetine rodents, first appeared
in the Old World in the Oligocene era (about
38 million years ago) and are found in North
America by the mid-Oligocene (about 33
million years ago). They were adapted to
forest environments, but as land dried during the succeeding Miocene era (26–8 million years ago) some became more terrestrial in their habits and developed into forms
recognizable as those of modern New World
rats and mice, the Hesperomyinae. In the
course of their evolution they occupied
many habitats similar to those occupied by
their Old World counterparts. In North
America, for example, there are harvest
mice of the genus *Reithrodontomys* that have
counterparts in the Old World in the mice of
the genus *Micromys*. The North American
wood mice of the genus *Peromyscus* have
counterparts in the Murinae genus *Apodemus*, the wood mice of the Old World.

The rats and mice of South America
developed in a similar way. After the land
bridge between North and South America
was formed during the Pliocene era (7–3
million years ago) several stocks of primitive
North American cricetine rodents, probably
animals equipped for climbing and in other
ways adapted for living in forests, moved

into South America where they underwent
an extensive radiation, in the first instance
for occupying spacious grassland habitats.
Subsequently they occupied many habitats,
some of which are not found occupied by
rodents in other parts of the world. A
particularly interesting feature of this radiation in South America is a consequence of
the absence in much of the continent of
members of the orders Insectivora (insectivores) and Lagomorpha (rabbits, hares and
pikas). Many South American rats and mice
have evolved in their forms and activities so
that they somewhat resemble shrews, water
shrews, moles and rabbits: the South American species include *Blarinomys breviceps*,
the Shrew mouse, the genus *Notiomys*, the
mole mice, the genus *Rheomys*, the fish-
eating mice, and the genus *Reithrodon*, the
rabbit rat.

This account to some extent simplifies
a very complicated evolutionary history.
Phases of change in South America have
occurred sporadically, often in response to
changes in environment. For example, during the Pleistocene era (2 million–10,000
years ago) zones of forest expanded and
contracted, sometimes creating isolated
groups which then developed along different
lines. The resulting mass of species and
genera creates considerable problems for the
natural historian trying to classify all the
animals belonging to the New World rats
and mice.

Though New World rats and mice include
a vast range of forms they are all small: the
largest living species are no longer, in head-
body length, than 1ft (about 30cm). Their
tails vary in length according to the extent of
adaptation for arboreal life. Highly arboreal
animals have long tails (usually they are
longer than half the animal's length), terrestrial forms the shortest. Most forms have a
brown back and white belly, but some
depart from this scheme and exhibit very
attractive coat colors, for example the Chinchilla mouse which has a strongly contrasting combination of buff to gray back and
white belly. The tails of most species have
few hairs, but the exceptional species have
well-haired tails, for example the Bushy-
tailed woodrat. All exhibit the basic arrangement of teeth common to highly
evolved rodents, viz. three molars on each
side of the jaw separated, by a distinct gap,
from a pair of incisors. The incisors grow
continuously. They have enamel on their
anterior surfaces which enables a sharp
cutting edge to be maintained.

In adapting for life in different environments New World rats and mice have

▲ **The unmatched deer mouse.** Of the 49 species of deer or white-footed mice of North and Central America *Peromyscus maniculatus* is most widespread. It can be found in alpine areas, forests, woodlands, grasslands and desert—anywhere except in moist habitats.

▶ **Rubbery droplets.** Mice reproduce rapidly, by a combination of short estrous cycle, short gestation and long breeding season. The average litter of mice is larger than the average mammal litter. Female deer mice produce on average 3.4 young per litter, in this as it were premature state, after a gestation of about three weeks. A similar period is required for weaning.

developed several modifications to the basic rat or mouse appearance. Burrowing forms have short necks, short ears, short tails and long claws. Forms adapted for an aquatic life often have webbed feet (for example the marsh rats of the genus *Holochilus*) or a fringe of hair on the hind feet which increases the surface areas of each foot (for example the fish-eating rats, genus *Daptomys*). In forms even more developed for aquatic life the external ear is reduced in size or even absent (as in the fish-eating rats of the genus *Anotomys*). Species that live in areas of semidesert or desert often have a light-colored back, and elsewhere the color of the back is often a shade that matches the surrounding soil. This enables exposed terrestrial New World rats and mice to reduce the threat they face from such predators as owls.

Above the generic level the classification of New World rats and mice is controversial, and subject to change. Genera can be combined to form 13 tribes, which in turn can be thought of as belonging to six groups, though the groups are of varying validity (see table).

The **White-footed mice** and their allies (tribe Peromyscini) consists of 10 genera, the genus *Peromyscus* containing the most species which are popularly known as white-footed mice or deer mice. In general they are nocturnal and eat seeds. Within the genus species show an array of adaptations. Head-body sizes range from 3–6.7in (8–17cm) and tail lengths from 1.6–8in (4–20.5cm). In terrestrial forms tails are shorter than the head-body length, in arboreal forms they are longer. Those that inhabit environments with great climatic fluctuations often produce many young. Species that occupy more stable habitats generally have low litter sizes and increased longevity, and also a relatively large brain. Such species include the California mouse, with a mean litter size of two and a brain weight equaling 2.9 percent of the body weight. An example of a species with a small brain and large average litter size is the Deer mouse, with a mean litter size of five and a brain weight that is 2.4 percent of the body weight.

Where several species of *Peromyscus* co-exist the habitat is subdivided in such a way that there is a segregation with respect of microhabitat. When three species co-exist, forming a guild, they are usually of three different sizes, with a small member having more versatile feeding habits and microhabitat requirements and often producing many young. The medium-sized member of the

guild is usually somewhat more restricted in its habitat requirements and presumably takes a narrower range of food sizes. The largest member of the guild will show the lowest fecundity, the greatest longevity, will have the most restricted microhabitat requirements, and is often specialized for feeding on larger seeds, nuts and fruits. A typical example is the three species complex in California: *P. truei*, *P. maniculatus* and *P. californicus*.

Closely related to the genus *Peromyscus* is the Volcano mouse which is a burrow user and is quite terrestrial in its habits. It occurs at elevations of 8,530 to 14,100ft (2,600m–4,300m) and there is a birth peak in July and August.

▲ **A home in the desert.** Desert woodrats that have access to rocks sometimes prefer to make their homes in crevices rather than in specially built nests. Woodrats are solitary: each has to keep watch for himself.

▶ **The Bushy-tailed woodrat,** whose range reaches into northwest Canada, is the northernmost of the woodrats. This may account for the animal's total insulation.

THE 6 GROUPS AND 13 TRIBES OF NEW WORLD RATS AND MICE

North American Neotomine-Peromyscine group

White-footed mice and their allies
Tribe Peromyscini.
Genera: **white-footed mice** or **deer mice**
(*Peromyscus*), 49 species, from N Canada (except high Arctic) S through Mexico to Panama; species include **California mouse** (*P. californicus*), **White-footed mouse** (*P. Leucopus*). **Harvest mice** (*Reithrodontomys*), 19 species, from W Canada and USA S through Mexico to W Panama; species include **Western harvest mouse** (*R. megalotis*). *Habromys*, 4 species, from C Mexico S to El Salvador. *Podomys floridanus*, Florida peninsula. **Volcano mouse** (*Neotomodon alstoni*), montane areas of C Mexico. **Grasshopper mice** (*Onychomys*), 3 species, SW Canada, NW USA S to N C Mexico. *Osgoodomys bandaranus*, W C Mexico. *Isthmomys*, 2 species, Panama. *Megadontomys thomasi*, C Mexico. **Golden mouse** (*Ochrotomys nuttalli*), SW USA.

Woodrats and their allies
Tribe Neotomini.
Genera: **woodrats** (*Neotoma*), 19 species, USA to C Mexico. **Allen's woodrat** (*Hodomys alleni*), W C Mexico. **Magdalena rat** (*Xenomys nelsoni*), W C Mexico. **Diminutive woodrat** (*Nelsonia neotomodon*), C Mexico.

Pygmy mice and brown mice
Tribe Baiomyini.
Genera: **pygmy mice** (*Baiomys*), 2 species, from SW USA S to Nicaragua. **Brown mice** (*Scotinomys*), 2 species, Brazil, Bolivia, Argentina.

Central American climbing rats
Tribe Tylomyini.
Genera: **Central American climbing rats** (*Tylomys*), 7 species, S Mexico to E Panama. **Big-eared climbing rat** (*Ototylomys phyllotis*), Yucatan peninsula of Mexico S to Costa Rica.

Nyctomyine group

Vesper rats
Tribe Nyctomyini.
Genera: **Central American vesper rat** (*Nyctomys sumichrasti*), S Mexico S to C Panama. **Yucatan vesper rat** (*Otonyctomys hatti*), Yucatan peninsula of Mexico and adjoining areas of Mexico and Guatemala.

Thomasomyine-Oryzomyine group

Paramo rats and their relatives
Tribe Thomasomyini.
Genera: **paramo rats** (*Thomasomys*), 25 species, areas of high altitude from Colombia and Venezuela S to S Brazil and NE Argentina. **South American climbing rats** (*Rhipidomys*), 7 species, low elevations from extreme E Panama S across northern S America to C Brazil. **Colombian forest mouse** (*Chilomys instans*), high elevations in Andes in W Venezuela S to Colombia and Ecuador. *Aepomys*, 2 species, high elevations in Andes in Venezuela, Colombia, Ecuador. **Rio de Janeiro rice rat** (*Phaenomys ferrugineus*), vicinity of Rio de Janeiro.

Rice rats and their allies
Tribe Oryzomini.
Genera: **rice rats** (*Oryzomys*), 57 species, SE USA S through C America and N S America to Bolivia and C Brazil. **Brazilian spiny rat** (*Abrawayaomys ruschii*), SE Brazil. **Galapagos rice rats** (*Nesoryzomys*), 5 species, Galapagos archipelago of Ecuador. **Giant rice rats** (*Megalomys*), 2 species (recently extinct), West Indies (Martinique and Santa Lucia). **Ecuadorean spiny mouse** (*Scolomys melanops*), Ecuador. **False rice rats** (*Pseudoryzomys*), 2 species, Bolivia, E Brazil, N Argentina. **Red-nosed mouse** (*Wiedomys pyrrhorhinos*), E Brazil. **Bristly mice** (*Neacomys*), 3 species, E Panama across lowland S America to N Brazil. **South American water rats** (*Nectomys*), 2 species, lowland S America to NE Argentina. **Brazilian arboreal mouse** (*Rhagomys rufescens*), SE Brazil.

Akodontine-Oxymycterine group

South American field mice
Tribe Akodontini.
Genera: **South American field mice** (*Akodon*), 33 species, found in most of S America (from W Colombia to Argentina). *Bolomys*, 6 species, montane areas of SE Peru S to Paraguay and C Argentina. *Microxus*, 3 species, montane areas of Colombia, Venezuela, Ecuador, Peru. **Cane mice** (*Zygodontomys*), 3 species, Costa Rica and N S America.

Burrowing mice and their relatives
Tribe Oxymycterini.
Genera: **burrowing mice** (*Oxymycterus*), 9 species, SE Peru, W Bolivia E over much of Brazil and S to N Argentina. **Andean rat** (*Lexonus apicalis*), SE Peru and W Bolivia. **Shrew mouse** (*Blarinomys breviceps*), E C Brazil. **Mole mice** (*Notiomys*), 6 species, Argentina and Chile. **Mount Roraima mouse** (*Podoxymys roraimae*), at junction of Brazil, Venezuela, Guyana. *Juscelimomys candango*, vicinity of Brasilia.

Ichthyomyine group

Fish-eating rats and mice
Tribe Ichthyomyini.
Genera: **fish-eating rats** (*Ichthyomys*), 3 species, premontane habitats of Venezuela, Ecuador, Peru. *Daptomys*, 3 species, French Guiana, premontane areas of Peru and Venezuela. **Water mice** (*Rheomys*), 5 species, C Mexico S to Panama. **Ecuadorian fish-eating rat** (*Neusticomys monticolus*), Andes region of S Colombia and N Ecuador.

Sigmodontine-Phyllotine-Scapteromyine group

Cotton rats and marsh rats
Tribe Sigmodontini.
Genera: **marsh rats** (*Holochilus*), 4 species, most of lowland S America. **Cotton rats** (*Sigmodon*), 8 species in S USA, Mexico, C America, NE S America as far S as NE Brazil; species include **Hispid cotton rat** (*S. hispidus*).

Leaf-eared mice and their allies
Tribe Phyllotini.
Genera: *Graomys*, 3 species, Andes of Bolivia S to N Argentina and Paraguay. *Andalgalomys*, 2 species, Paraguay and NE Argentina. *Galenomys garleppi*, high altitudes in S Peru, W Bolivia, N Chile. *Auliscomys*, 4 species, mountains of Bolivia, Peru, Chile and Argentina. **Puna mouse** (*Punomys lamminus*), montane areas of S Peru. **Rabbit rat** (*Reithrodon physodes*), steppe and grasslands of Chile, Argentina, Uruguay. **Vesper mice** (*Calomys*), 8 species, most of lowland S America. **Chinchilla mouse** (*Chinchillula sahamae*), high elevations S Peru, W Bolivia, N Chile, Argentina. **Chilean rat** (*Irenomys tarsalis*), N Argentina, N Chile. **Andean mouse** (*Andinomys edax*), S Peru, N Chile. **Highland desert mouse** (*Eligmodontia typus*), S Peru, N Chile, Argentina. **Leaf-eared mice** (*Phyllotis*), 12 species, from NW Peru S to N Argentina and C Chile. **Patagonian chinchilla mice** (*Euneomys*), 4 species, temperate Chile and Argentina. **Andean swamp rat** (*Neotomys ebriosus*), Peru S to NW Argentina.

Southern water rats and their allies
Tribe Scapteromyini.
Genera: **red-nosed rats** (*Bibimys*), 3 species, SE Brazil W to NW Argentina. **Argentinean water rat** (*Scapteromys tumidus*), SE Brazil, Paraguay, E Argentina. **Giant South American water rats** (*Kunsia*), N Argentina, Bolivia, SE Brazil.

The size of harvest mice varies considerably from 2–5.7in (6–14cm) in head-body length, from 2.5–3.7in (6.5–9.5cm) in tail length. The North American species tend to be smaller than the Central American species, rarely weighing more than 0.5oz (15g). The Western harvest mouse is typical of the grassland areas of western North America. A nocturnal seed or grain eater it constructs globular nests approximately 9in (24cm) off the ground in tall grass.

Grasshopper mice are specialized forms that inhabit arid and semiarid habitats. In size they range from 3.5–5.1in (9–13cm) in head-body length, though their tail is rather short, 1.2–2.4in (3–6cm). These mice feed on insects and small vertebrates. They construct burrows in which they live as pairs during the breeding season. Litters disperse upon weaning and the duration of pairing in subsequent seasons is unknown. These rodents are well known for their high-pitched squeak (usually above 20kHz) which may function as a spacing call. Both sexes emit these calls, but males call more frequently than females. In nature the burrows are widely spaced.

The Golden mouse is confined to the moderately wet wooded habitats of the southeastern USA. The distinctive golden brown color of its back contrasts sharply with the white belly. This is an extremely arboreal form which builds a complex leafy nest in tangles of vines.

The **woodrats** and their allies (tribe Neotomini) are rat-sized rodents, varying in color from dark buff on the back to paler shades on the belly. In general they eat a wide range of foods but some species are highly adapted for feeding on the green parts of plants; indeed, *Neotoma stephensi* feeds almost entirely on the foliage of juniper trees. Many species are adapted for living in and around crevices or cracks in rocky outcrops, others construct burrows. But whether burrowers or rock-dwellers all have the habit of creating mounds of sticks and other detritus in the vicinity of the nest hole or crack. Desert species often utilize such items as pads of spiny cacti. This habit and the transportation of materials to the mound have earned them the name "pack rats" in some parts of their range. Each stick nest tends to be inhabited by a single adult individual, but individuals may visit neighboring nests; in particular males apparently visit females when they are receptive.

The Magdalena rat occurs in an extremely restricted area of tropical deciduous forests in western Mexico in the states of Jalisco and Colima, where it may have an extended season of reproduction. It is a small nocturnal woodrat with excellent climbing ability.

The Diminutive woodrat is found in the mountainous areas of central and western Mexico where it is known to shelter in crevices of rock outcroppings at elevations exceeding 6,500ft (2,000m).

Pygmy mice and **brown mice** (tribe Baiomyini) are the smallest New World rodents. Pygmy mice have a relatively small home range (often less than 9,700sq ft, about 900sq m) compared with a larger seed-eating rodent such as *Peromyscus leucopus*, which has a home range of 2.9–6.9 acres (1.2–2.8ha). Pygmy mice are seedeaters which inhabit a grass nest, usually under a stone or log. They may be monogamous while pairing and rearing their young.

Brown mice are small subtropical mice which employ a high-pitched call apparently to demarcate territory. Males produce this call more frequently than females.

The **Central American climbing rats** (tribe Tylomyini; two genera) are associated with water edge forested habitat at lower elevations. They are extremely arboreal, strictly nocturnal and feed primarily on fruits, seeds and nuts. The **Big-eared climbing rat** is smal-

▲ **Representatives from six tribes** of New World rats and mice. (**1**) A South American climbing rat (genus *Rhipidomys*; tribe Thomasomyini). (**2**) A Central American vesper rat (genus *Nyctomys*; tribe Nyctomini). (**3**) A Central American climbing rat (genus *Tylomys*; tribe Tylomyini). (**4**) A pygmy mouse (genus *Baiomys*; tribe Baiomyini). (**5**) A white-footed or deer mouse (genus *Peromyscus*; tribe Peromyscini). (**6**) A woodrat or pack rat, carrying a bone (genus *Neotoma*; tribe Neotomini).

ler than the other species and again is also confined to lowland tropical forests. Although adapted for living in trees it also forages on the ground. It produces small litters of fully haired young whose eyes open after about six days. These well-developed young are a unique departure from the normal pattern found in New World rats and mice which usually produce hairless young whose eyes open after ten to twelve days. The gestation of the Big-eared climbing rat is approximately 50 days. Most New World rats and mice have gestation periods of approximately three weeks.

The tribe Nyctomyini contains just two species of **vesper mice**. The Central American vesper rat is a specialized nocturnal, arboreal fruit-eater which builds nests in trees and has a long tail and large eyes.

The Yucatan vesper rat, also highly arboreal, is a relict species in the Yucatan peninsula. It probably was once more broadly distributed under different climatic conditions but became isolated in the Yucatan during one of the drying cycles in the Pleistocene era (2 million–20,000 years ago).

Paramo rats and their allies (tribe Thomasomyini) are distributed throughout South America. In the Andes many species of paramo rats are adapted for life at elevations exceeding 13,000ft (4,000m). Otherwise they are almost always confined to forests or to forests along rivers. They are nocturnal and fruit-eating, but their biology is very poorly known. Litters of two to four young have been recorded, but in general their reproductive potential is considered to be quite low.

The South American climbing rats are likewise adapted for life in trees. They are also nocturnal, and feed upon fruits, seeds,

fungi and insects. Their litter size is small: for *R. mastacalis* two or three young per litter have been recorded.

Rice rats and their allies (tribe Oryzomini) are an assemblage with two tendencies: a retention of the long tail as an adaptation for a life spent partly in trees or, alternatively, the exploitation of moist habitats, culminating in semiaquatic adaptations. Many of the species are adapted to varying altitudes. For example, in northern Venezuela *Oryzomys albigularis* occurs above elevations of 3,300ft (about 1,000m); at lower elevations it is replaced by *O. capito*. When two species of rice rats occur in the same habitat one is often more adapted to life in trees than the other. Such is the case when *O. capito*, a terrestrial species, occurs with *O. bicolor*, a species adapted for climbing.

The South American water rat (one of the two species of *Nectomys*) is semiaquatic and the dominant aquatic rice rat over much of South America. *Oryzomys palustris*, found on the gulf coast of northern Mexico north to the coast of southern Maryland (USA), is also semiaquatic in its habits. This rice rat has a wide-ranging diet though at certain times of the year over 40 percent of its food may consist of snails and crustaceans. Across much of its range it has an extended breeding season from February to November, and is able to produce four young at 30-day intervals with the result that it can rapidly become a serious agricultural pest.

The small bristly mice or spiny rice rats, which have a distinctive spiny coat, are nocturnal and eat seeds. *Neacomys tenuipes*, in northern South America, can exhibit wide variations in population density.

Rice rats have excelled at colonizing islands in the Caribbean and the Galapagos Islands. Many island species of *Oryzomys* are currently threatened with extinction, due to human activities. The introduction of the Domestic house cat and murine rats and mice has had a severe impact on the Galapagos rice rats.

South American field mice (tribe Akodontini) are adapted for foraging on the ground and many are also excellent burrowers.

Akodon species have radiated to fill a variety of habitats. In general the species are omnivorous, including in their diets green vegetation, fruits, insects and seeds. Most species of *Akodon* are adapted to moderate to high elevations. *Akodon urichi* typifies the adaptability of the genus. Since it tends to be active both in daytime and at night it is terrestrial and eats an array of food items including fruits, seeds and insects.

Members of the genus *Bolomys* are closely allied to *Akodon* but are more specialized in adaptations for terrestrial existence. They have short ears, a short tail and a body form very similar to that of the Field vole. Members of the genus *Microxus* are similar to *Bolomys* in appearance but their eyes show even further reduction in size. They are strongly adapted for a terrestrial burrowing life as evidenced by a short neck, reduced external ear length, and reduction in eye size.

Cane mice are widely distributed in South America, taking the place of *Akodon* at low elevations in grasslands and bushlands. In grasslands it sometimes constructs runways which can be visible to the human observer. Cane mice eat a considerable quantity of seeds and do not seem to be specialized for processing green plant food. In grassland habitats subject to seasonal fluctuations in rainfall the cane mice may show vast oscillations in population density. In the llanos of Venezuela they show population explosions when productivity of the grasslands is exceptionally high, enabling them to harvest seeds and increase their production of young. Densities can vary from year to year from a high of 6 per acre (15 per ha) to a low of less than 0.4 per acre.

The **burrowing mice** and their relatives (tribe Oxymycterini) are closely allied to the South American field mice. They are distinguished by adaptations for burrowing habits and a trend within the species assemblage towards specialization for a more insectivorous food niche. Rodents specialized in this way show a reduction in the size of their molar teeth, and an elongate snout. Longer claws aid in excavating the soil for soil arthropods, larvae and termites.

Burrowing mice of the genus *Oxymycterus* are long-clawed with a short tail. Their molar teeth are weak, their snout long. These features correlate with eating insects. A grass nest is made in a burrow system where the young, two or three, are born. The small litter size of some species of

▼ **Representatives of seven tribes** of New World rats and mice. (1) A South American field mouse (genus *Akodon*; tribe Akodontini) grooming its tail. (2) A fish-eating rat (genus *Ichthyomys*; tribe Ichthyomyini). (3) The Argentinian water rat (*Scapteromys tumidus*; tribe Scapteromyini). (4) A cotton rat (genus *Sigmodon*; tribe Sigmodontini) attempting to remove an egg. (5) A mole mouse (genus *Notiomys*; tribe Oxymycterini). (6) A South American water rat (genus *Nectomys*; tribe Oryzomini). (7) A leaf-eared mouse (genus *Phyllotis*; tribe Phyllotini).

Oxymycterus may correlate with a lowered metabolic rate as an adaptation to termite feeding. Mammals specializing for feeding on termites often have a metabolic rate lower than would be expected, probably because the high chitin content of insects gives a lower return in net energy than other protein and carbohydrates. Lower metabolic rates in termite- and ant-feeding forms is an outcome of adjustment to the rate of energy return. A side effect of this is a reduced reproductive capacity, reflected in smaller litter sizes.

The Shrew mouse (one species), represents an extreme adaptation for a burrowing way of life. Its eyes are reduced in size and its ears are so short that they are hidden in the fur. It burrows under the litter of the forest floor and can construct a deep, sheltering burrow. Its molar teeth are very reduced in size, a characteristic that indicates adaptation for a diet of insects.

Mole mice are widely distributed in Argentina and Chile and exhibit an array of adaptations reflected in their exploitation of both semiarid steppes and wet forests. Some species are adapted to higher elevation forests, others to moderate elevations in central Argentina. They have extremely powerful claws which may exceed 0.3in (0.7cm) in length. The name mole mice derives from their habits of burrowing and spending most of their life underground.

The **fish-eating rats and mice** (tribe Ichthyomyini) represent an interesting group in that they have specialized for a semiaquatic life and have altered their diets so as to feed on aquatic insects, crustaceans, mollusks and fish.

Little is known of the details of the biology of fish-eating rats and mice other than that they occur on or near high-elevation fresh water streams and exploit small crustaceans, arthropods and fish as their primary food sources.

Fish-eating rats of the genus *Ichthyomys* are among the most specialized of the genera. They have a head-body length that almost attains 9in (33cm), which is just exceeded by the tail. Their fur is short and thick, their eyes and ears are reduced in size, and their whiskers are stout. A fringe of hairs on the toes of the hind feet aids in swimming, and the toes are partially webbed.

They resemble a large water shrew or some of the fish-eating insectivores of West Africa and Madagascar.

Fish-eating rats of the genus *Daptomys* are similar to *Ichthyomys* and are distributed disjunctly in the mountain regions of Venezuela and Peru. The Ecuadorian fish-eating rat is the least specialized for aquatic life.

Water mice are slightly smaller than *Ichthyomys* and rarely exceed 7in (19cm) in head-body length. They occur in mountain streams of central America and Colombia and are known to feed on snails and possibly fish. In the webbing of their hind feet and in the hairs on the outer sides of their feet they are similar to *Ichthyomys*.

Cotton rats and **marsh rats** (tribe Sigmodontini) are united by a common feature, namely, folded patterns of enamel on the molars which when viewed from above tend to approximate to an "S" shape. They exhibit a range of adaptations; the species referred to as marsh rats are adapted for a semiaquatic life, whereas the cotton rats are terrestrial. Both groups, however, feed predominantly on herbaceous vegetation.

The genus *Holochilus* contains the web-footed marsh rats, of which two species (of four) are broadly distributed in South America. The underside of the tail has a fringe of hair, an adaptation to swimming. They build a grass nest near water, sufficiently high to prevent flooding, which may exceed 15.7in (40cm) in diameter. In the more southern parts of their range in temperate South America breeding tends to be confined to the spring and summer (ie September–December).

Cotton rats are broadly distributed from the southern USA to northern South America. In line with their adaptations for terrestrial life the tail is always considerably shorter than head-body length. Cotton rats are active during both day and night, and although they eat a wide range of food they consume a significant amount of herbs and grasses during the early phases of vegetational growth after the onset of rains. A striking feature of the reproduction of the Hispid cotton rat is that its young are born fully furred and their eyes open within 36 hours of their birth. It has a very high reproductive capacity and although it produces well-developed (precocial) young, the gestation period is only 27 days. The litter size is quite high, from five to eight, with 7.6 as an average. The female is receptive after giving birth and only lactates for approximately 10–15 days. Thus the turn-around time is very brief and a female can produce a litter every month during the breeding season. In agricultural regions this rat can become a serious pest.

Leaf-eared mice and their allies (tribe Phyllotini) are typified by the genera *Phyllotis* and *Calomys*. *Calomys* (vesper mice) includes a variety of species distributed over most of South America. They have large ears (as do *Phyllotis*) and feed primarily on

▲ **Equilibrium on a twig.** American harvest mice are nimble, agile rodents, adept at climbing. Their nests are usually found above ground, in grasses, low shrubs, or small trees: they are globular in shape, woven of grass.

▶ **A wolf among mice.** All three insect-eating species of grasshopper mice stand erect to utter repeated shrill sounds, each lasting about a second. They occur when one mouse detects another nearby, or just as a mouse is about to kill.

◀ **A place in the sun.** The leaf-eared mouse of northwest South America is, unlike many mice, active by day. It is often found out on rocks, standing in sunlight, keeping company with viscachas.

plant material; arthropods are an insignificant portion of their diet. The genus *Phyllotis* (leaf-eared mice) is composed of several species, most of which occur at high altitudes. They are often active in the day and may bask in the sun. They feed primarily on seeds and herbaceous plant material. The variation in form and the way in which several species of different size occur in the same habitat are reminiscent of *Peromyscus*, the deer mice (tribe Peromyscini).

The Rabbit rat is of moderate size and has thick fur adapted to the open country plains of temperate Chile, Argentina and Uruguay. It feeds primarily on herbaceous plant material and may be active throughout both day and night.

The Highland desert mouse is one of the few South American rodents specialized for semiarid habitats. Its hind feet are long and slender, resulting in a peculiar gait where the forelimbs simultaneously strike the ground followed by a power thrust from the hind legs where the forelimbs leave the ground. The kidneys of this species are very efficient at recovering water; indeed, it can exist for considerable periods of time without drinking, being able to derive its water as a by-product of its own metabolism.

Patagonian chinchilla mice are distributed in wooded areas, from central Argentina south to Cape Horn. The Puna mouse is found only in the altiplano of Peru. This rodent is the most vole-like in body form of any South American rodent. It is active both day and night and its diet is apparently confined to herbaceous vegetation. The Chilean rat is an inhabitant of humid temperate forests. This is an extremely arboreal species. It may be a link between the phyllotines and the oryzomyine rodents or rice rats.

The Andean marsh rat occurs at high elevations near streams and appears to occupy a niche appropriate for a vole.

The **southern water rats** and their allies (tribe Scapteromyini) are adapted for burrowing in habitats by or near rivers. The Argentinean water rat is found near rivers, streams and marshes. It has extremely long claws and can construct extensive burrow systems.

The giant South American water rats prefer moist habitats and have considerable burrowing ability. *Kunsia tomentosus* is the largest living New World rat with a head-body length that may reach 11in (28cm) and a tail length of up to 6.3in (16cm).

Red-nosed rats are small burrowing forms allied to the larger genera, but whose biology and habits are poorly known.   JFE

VOLES AND LEMMINGS

Subfamily: Microtinae
One hundred and ten species in 18 genera.
Family: Muridae.
Distribution: N and C America, Eurasia, from
Arctic S to Himalayas, small relic population in
N Africa.

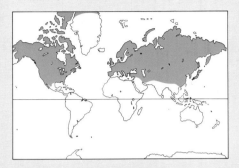

Habitat: burrowing species are common in
tundra, temperate grasslands, steppe, scrub,
open forest, rocks; 5 species are aquatic or
arboreal.

Size: most species head-body
length 4–4.5in (10–11cm),
tail length 1.2–1.6in
(3–4cm), weight 0.6–0.7oz
(17–20g).

Gestation: 16–28 days.

Longevity: 1–2 years.

▶ **On the nibble:** a Bank vole. Bank voles live
in Europe and Central Asia in woods and
scrubs, in banks and swamps, usually within a
home range of about 2 acres (0.8ha).

▶ **A summer scene in the arctic tundra.**
Collared lemmings (including these Arctic
lemmings) live along the arctic edge of
northern Europe, Siberia and North America.
In summer, coats are brown and heavy, but
with the approach of winter, coats turn white
and on the front paws the third and fourth
claws grow a second prong.

THE fascinating lemmings and voles that
belong to the Muridae subfamily Microtinae have two features of particular interest.
Firstly their populations expand and contract considerably, in line with cyclical
patterns. This has made them the most
studied subfamily of rodents (and the basis
of much of our understanding about the
population dynamics of small mammals).
Secondly, though they neither hibernate
like such larger mammals as ground squirrels, nor can rely on a thick layer of fat like
bears, most voles and lemmings live in
habitats covered by snow for much of the
year. They are able to survive thanks to their
ability to tunnel beneath the snow where
they are insulated from extreme cold.

Vole and lemmings are small, thickset
rodents with bluntly rounded muzzles and
tails usually less than half the length of their
bodies. Only small sections of their limbs are
visible. Their eyes and ears tend to be small
and in lemmings the tail is usually very

short. Coat colors vary not only between
species but often within them. Lemmings'
coats are especially adapted for cold temperatures: they are long, thick and waterproof.
The Collared lemming is the only rodent
that molts to a complete white coat in
winter.

Some species display special anatomical
features: the claws of the first digit of the
Norway lemming are flattened and enlarged
for digging in the snow while each fall the
Collared lemming grows an extra big claw
on the third and fourth digits of its forelegs
and sheds them in spring. Muskrats have
long tails and small webbing between toes
which assist in swimming. The mole lemmings, adapted for digging, have a more
cylindrical shape than other species and
their incisors, used for excavating, protrude
extremely.

Adult males and females are usually the
same color and approximately the same size,
though the color of juveniles' coats may

differ from the adults'. Although most adult voles weigh less than 3.5oz (100g) the muskrat grows to over 50oz (1,400g). The size of the brain, in relation to body size, is lower than average for mammals.

Smell and hearing are important, well-developed senses, able to respond, respectively, to the secretions that are used to mark territory boundaries, indicate social status and perhaps to aid species recognition, and to vocalizations (each species has a characteristic suite of calls). Calls can be used for alarm, to threaten or as part of courtship and mating. Brandt's voles, which live in large colonies, sit up on the surface and whistle like prairie dogs.

Microtines, especially lemmings, are widely distributed in the tundra regions of the northern hemisphere where they are the dominant small mammal species. Their presence there is the result of recolonization since the retreat of the last glaciers. Voles are also found in temperate grasslands and in the forests of North America and Eurasia.

Because the Pleistocene era (2 million-10,000 years ago) has yielded a rich fossil record of microtine skulls, much of the taxonomy of this subfamily is based on the kind and structure of teeth. They are also used to distinguish the microtines from other rodents. All microtine teeth have flattened crowns and prisms of dentine surrounded by enamel. There are 12 molars (3 on each side of the upper and lower jaws) and 4 incisors. Species are differentiated by the particular pattern of the enamel of the molars. Dentition has not proved sufficient

for solving all taxonomic questions, and some difficulties remain both in delimitating species and defining genera. It is often possible to distinguish the species of live specimens by using general body size, coat color or length of tail. Subdivision of species into many subspecies (not listed here) reflects geographic variation in coat color and size in widely distributed species. *Microtus*, the largest genus (accounting for nearly 50 percent of the subfamily), is a heterogeneous collection of species.

Many species have widely overlapping ranges. For example, there are six species in the southern tundra and forest of the Yamal Peninsula in the USSR: two lemmings and four voles. Each can be differentiated by its habitat preference or diet. The ranges of Siberian lemmings and Collared lemmings overlap extensively but Collared lemmings prefer upland heaths and higher and drier tundra whereas Siberian lemmings are found in the wetter grass-sedge lowlands. In the northwest USA the Long-tailed vole is actively excluded from grasslands by the dominant Montane vole.

Voles and lemmings are herbivores, and usually eat the green parts of plants, but some species prefer bulbs, roots, mosses or even pine needles. The muskrat occasionally eats mussels and snails. Diet usually changes with the seasons and varies according to location, reflecting local abundances of plants. Species living in moist habitats, such as lemmings and the Tundra vole, prefer grasses and sedges while those species in drier habitats such as the Collared lemmings prefer herbs. But animals select their food to some extent; diets do not just simply mimic vegetation composition.

Voles and lemmings can be found foraging both day and night, although dawn and dusk might be preferred. They obtain food by grazing, or digging for roots; grass is often clipped and placed in piles in their runways. Some cache food in summer and fall, but in winter, when the snow cover insulates the animals that nest underneath the surface, food is obtained by burrowing and animals also feed on plants at ground surface. The Northern red-backed vole feeds on, among other items, berries, so in summer has to compete for them with birds. In winter, when the bushes are covered with snow, this vole can burrow to reach the berries. So only during the spring thaw is the animal critically short of food. In winter the Sagebrush vole utilizes the height of snow packs to forage on shrubs it normally cannot reach.

Most microtines, for example all species of

microtus, the Florida water rat and the Steppe lemming, have continuously growing molars and can chew more abrasive grasses and a greater quantity of grass than species with rooted molars. In tundra and grasslands voles and lemmings help the recycling of nutrients by eliminating waste products and clipping and then storing food below ground.

The life span of microtines is short. They reach sexual maturity at an early age and are very fertile. Mortality rates, however, are high: during the breeding season, of the animals alive one month only 70 percent are alive the next. The age of sexual maturity can vary considerably. The females of some species may become sexually mature only two or three weeks from birth. Males take longer, usually six to eight weeks. The Common vole has an extraordinarily fast development time. Females have been observed coming into heat while still suckling and may be only five weeks old at the birth of

their first littler. In species that breed only during the summer, young born early will probably breed that summer while later litters may not become sexually mature until the following spring.

The length of the breeding season is highly variable but lemmings can breed in both summer and winter. Winter breeding is less common in voles, which tend to breed from late spring to fall. Voles in Mediterranean climates, on the other hand, breed in winter and spring, during the wet season, but not in summer. The breeding season may vary within species in different years or in different parts of the range. The muskrat will breed all year in the southern part of its range but only in summer elsewhere. Some species of meadow voles will breed in winter during the phase of population increase but not when numbers are declining.

In many species, such as the Montane, Field and Mexican voles, the presence of a male will induce ovulation in a female.

▶ ▼ **Representative species of voles and lemmings.** (1) A muskrat (*Ondata zibethicus*) sitting on its house of branches and twigs. (2) A Red-tree vole (*Phenacomys longicaudatus*), a highly specialized, tree-living vole. (3) A Southern (or Afghan) mole-vole (*Ellobius fuscocapillus*). (4) A Taiga vole (*Microtus xanothognathus*). (5) A Norway lemming (*Lemmus lemmus*). (6) A Meadow vole (*Microtus pennsylvanicus*), drumming an alarm signal with its hind foot. (7) A Collared (or Arctic) lemming (*Dicrostonyx torquatus*) in its winter coat with double foreclaws. (8) A Collared lemming wearing its summer coat with single foreclaws. (9) A European water vole (*Arvicola terrestris*).

The 3 tribes of voles and lemmings

Lemmings
Tribe Lemmini.
Four genera and 9 species in N America and Eurasia inhabiting tundra, taiga and spruce woods. Skull broad and massive, tail very short, hair long; 8 mammae. Genera:

Collared lemmings
Genus *Dicrostonyx*, 2 species including: **Collared** or **Arctic lemming** (*D. torquatus*).

Brown lemmings
Genus *Lemmus*, 4 species including: **Norway lemming** (*L. lemmus*).

Wood lemming
Myopus schisticolor.

Bog lemmings
Genus *Synaptomys*, 2 species including: **Northern bog lemming** (*S. borealis*).

Mole lemmings
Tribe Ellobii.
One genus and 2 species in C Asia inhabiting steppe. Form is modified for a subterranean life; coat color varies from ocher sand to browns and blacks; tail short; no ears; incisors protrude forwards. Species: **Northern mole-vole** (*Ellobius talpinus*); **Southern** or **Afghan mole-vole** (*E. Fuscocapillus*).

Voles and mole-voles
Tribe Microtini.
Thirteen genera and 99 species in N America, Europe, Asia and the Arctic. Main genera are:

Red-backed and bank voles
Genus *Clethrionomys*.
Japan, N Eurasia, N America, inhabiting forest, scrub and tundra. Back usually red, cheek teeth rooted in adults, skull weak; 8 mammae; seven species including: **Bank vole** (*C. glareolus*), **Northern red-backed vole** (*C. rutilus*).

Meadow voles
Genus *Microtus*.
N America, Eurasia, N Africa. Coat and size highly variable, molars rootless, skull weak; 4–8 mammae; burrows on surface and underground. Forty-four species including: **Brandt's vole** (*M. brandti*), **Common vole** (*M. arvalis*), **Field vole** (*M. agrestis*), **Meadow vole** (*M. pennsylvanicus*), **Mexican vole** (*M. mexicanus*), **Montane vole** (*M. montanus*), **Prairie vole** (*M. ochrogaster*), **Taiga vole** (*M. xanothognathus*), **Townsend's vole** (*M. townsendii*), **Tundra vole** (*M. oeconomus*).

Pine voles
Genus *Pitymys*.
Not clearly distinct from *Microtus*. E and S USA, E Mexico, Eurasia. Fur soft, dense, mole-like; skull weak; 4–8 mammae; more given to digging than *Microtus*. Seventeen species including: **American pine vole** (*P. pinetorum*), **Mediterranean pine vole** (*P. duodecimcostatus*).

The other 10 genera of the tribe Microtini: **Mountain voles** (*Alticola*), 4 species in C Asia including: **Large-eared** or **High Mountain vole** (*A. macrotis*). **Water voles** (*Arvicola*), 3 species in N America and N Eurasia including: **European water vole** (*A. terrestris*). **Martino's snow vole** (*Dinaromys bogdanovi*), Yugoslavia. **Oriental voles** (*Eothenomys*), 11 species in E Asia including: **Père David's vole** (*E. melanogaster*). *Hyperacrius*, 2 species in Kashmir and the Punjab including: **True's vole** (*H. fertilis*). **Steppe lemmings** (*Lagarus*), 3 species in C Asia and W USA including: **Sagebrush vole** (*L. curtatus*). **Florida water rat** (*Neofiber alleni*), Florida. **Muskrat** (*Ondata zibethicus*), N America. *Phenacomys*, 4 species in W USA and Canada including: **Red-tree vole** (*P. longicaudus*). **Long-clawed mole-vole** (*Prometheomys schaposchnikowi*), Caucasus, USSR.

3

6

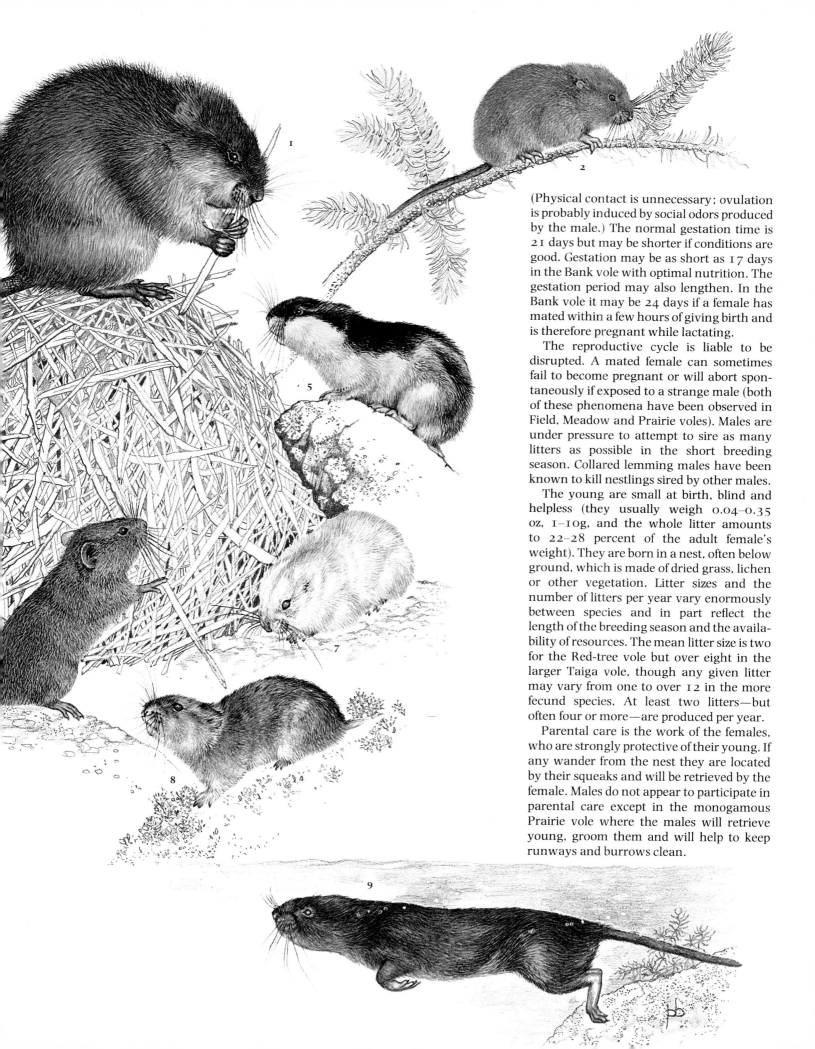

(Physical contact is unnecessary; ovulation is probably induced by social odors produced by the male.) The normal gestation time is 21 days but may be shorter if conditions are good. Gestation may be as short as 17 days in the Bank vole with optimal nutrition. The gestation period may also lengthen. In the Bank vole it may be 24 days if a female has mated within a few hours of giving birth and is therefore pregnant while lactating.

The reproductive cycle is liable to be disrupted. A mated female can sometimes fail to become pregnant or will abort spontaneously if exposed to a strange male (both of these phenomena have been observed in Field, Meadow and Prairie voles). Males are under pressure to attempt to sire as many litters as possible in the short breeding season. Collared lemming males have been known to kill nestlings sired by other males.

The young are small at birth, blind and helpless (they usually weigh 0.04–0.35 oz, 1–10g, and the whole litter amounts to 22–28 percent of the adult female's weight). They are born in a nest, often below ground, which is made of dried grass, lichen or other vegetation. Litter sizes and the number of litters per year vary enormously between species and in part reflect the length of the breeding season and the availability of resources. The mean litter size is two for the Red-tree vole but over eight in the larger Taiga vole, though any given litter may vary from one to over 12 in the more fecund species. At least two litters—but often four or more—are produced per year.

Parental care is the work of the females, who are strongly protective of their young. If any wander from the nest they are located by their squeaks and will be retrieved by the female. Males do not appear to participate in parental care except in the monogamous Prairie vole where the males will retrieve young, groom them and will help to keep runways and burrows clean.

While the young are living in the nest they learn to recognize the scent and behavioral cues of their species. So if young Brown lemmings are fostered (artificially) with Collared lemmings they will be more aggressive towards their own species as adults than to their fostered species. The length of lactation varies, but usually lasts about three weeks and in some species is terminated by the female abandoning the nest.

Until recently microtines were considered to have a relatively simple social system. A decade of careful experimentation and observation has shown, however, that the two sexes and different species have different social behaviors and that these may change radically between seasons.

The key to understanding both the social relationships and the spatial organization of voles and lemmings is the spacing behavior of males and females. In the breeding season males and females form territories delimited by scent marks. Male and female territories may be separate (as in the Montane vole and the Collared lemming), overlap (as in the European water and Meadow voles) or several females may live in overlapping ranges within a single male territory (for example, Taiga and Field voles). Males form

hierarchies of dominance; subordinates may be excluded from breeding. They may act to exclude strange males from an area—to reduce the incidence of induced abortion. The Common, Sagebrush and Brandt's voles live in colonies. The animals build complicated burrows.

Although in most species males are promiscuous, a few are monogamous. In the Prairie vole the males and females form pair bonds while the Montane vole appears to be monogamous at low density, as a result of the spacing of males and females, but with no pair bond. At high density this species is polygynous (males mating with several females). Monogamy is favored when both adults are needed to defend the breeding territory from intruders.

The social system may vary seasonally. In the Taiga vole the young animals disperse and their territories break down late in summer. Groups of 5–10 unrelated animals then build a communal nest which is occupied throughout the winter by both males and females. Communal nesting or local aggregations of individuals is also observed in the Meadow, Gray and Northern red-backed voles. Huddling together reduces the energy requirements.

Dispersal (the movement from place of

▶ **Territories of seven female water voles** along a stretch of water in the Bure Marshes (Norfolk, England), approximately 0.8mi (0.5km) long, in (a) January and (b) March. In January individuals occupied overlapping and undefended territories. However, by March the females had extended their ranges which now did not overlap with those of other individuals and were defended against intruders. One individual was unable to find a territory and migrated; another new individual came in to occupy a territory.

▶ **Swimming without trouble.** BELOW Many populations of water voles, distributed across Western Europe, are able to swim, both on the surface and underwater, even along the bottoms of streams. A few develop hairy fringes on their feet, but in general there seem to be no specific adaptations for aquatic life.

▼ **Evidence for population cycles** can come in two forms: (a) indirectly from records kept by the Hudson Bay Company for trade in Arctic fox pelts. The main prey of this predator is lemmings, so when prey densities increase so do those of the Arctic fox; (b) directly from the numbers of voles trapped in field experiments (the figures refer to the numbers of the Red-gray vole trapped in northern USSR over a sample period of 100 days). Both graphs show clearly cyclical changes in population levels taking three to four years.

Population Cycles in Lemmings and Voles

Populations of many species of lemmings and voles fluctuate in regular patterns, loosely called "cycles," in which high populations recur at intervals of 3–4 years. These cycles are most famous in lemmings on the tundra but they also occur in a great variety of species of voles.

The cycles are generated by changes in the rates of reproduction and mortality. Reproduction is at maximum rate during the phase of increase, augmented by a breeding season of increased length (often involving winter breeding in lemmings) and by animals achieving sexual maturity at a very young age. When the population reaches its peak reproduction is curtailed by a shortening of the breeding season and an increase in the age of sexual maturity. At the same time juvenile mortality rises and the population stops growing; a large number of animals die of stress and food shortage. The next year the population is very low and can make up for this only with three or four years of reproduction.

There are two competing schools of thought. One suggests that a population's tendency is to increase but cycles are induced because high-density populations exhaust the supply of high-quality food, and so population

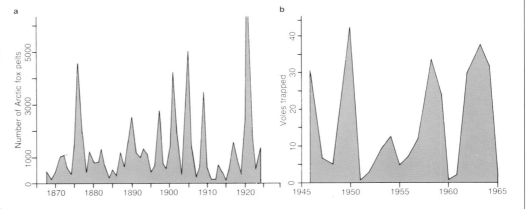

declines from malnutrition. Some herbs are toxic to voles. If a vole population exhausts its supply of palatable forage, individuals may be forced to feed on toxic plants or face starvation. An alternative view is that cycles arise because of changes of behavior within populations as individuals strive to increase their fitness in a fluctuating population. As populations grow individuals become more aggressive in defending their territories against intruders, and this increased aggression may inhibit reproduction and result in additional mortality from fighting or dispersal.

The two sexes often have different strategies. Males compete with other males for

access to reproductive females. Males may show infanticidal behavior. Females defend their nests from marauding individuals of both sexes and may also kill strange juveniles who wander into their territory. Intruder pressure increases with population density and at peak densities social strife can stop all reproduction and cause a decline in numbers.

Natural selection may thus favor large, aggressive individuals at high density because only these individuals will be able to hold a breeding territory under crowded conditions, and then favor small, docile individuals at low density when there is no one around to fight and the premium is on rapid reproduction.

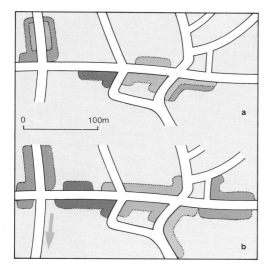

birth to place of breeding) is one of the most important aspects of microtine behavior and has been the subject of considerable research. It plays an important role in regulating population size and allows the animals to exploit efficiently the highly seasonal patchy nature of their habitat: strategies change according to population density. Norway lemmings move in summer into wet grassy meadows but in the fall move into deep mossy hillside habitats to over-winter (see pp72–3).

Dispersers differ from nondispersing animals in a number of ways. In the Meadow and Prairie voles animals that disperse on average are genetically different from permanent residents. Young males often disperse, forced to leave by the aggressive behavior of the dominant males. Additionally, in the Field, Prairie, Meadow, Tundra and European water voles many young pregnant females will disperse. This is quite unusual for mammals but it does enable voles to colonize rapidly vacant habitats.

The numbers of juvenile males and females are equal, except in the Wood lemming which has an unusual feature for mammals: some females are genetically programmed to have only daughters which is advantageous in an increasing population. In most species, however, there are usually more female adults than males, probably as a consequence of dispersal by males who are then more susceptible to predators. In the European water vole it is the females who usually disperse and also in this species there is a slight excess of males in the adult population. In Townsend's vole the survival and growth of juveniles are lower when the density of adult females is high.

Territory size varies but males usually have larger territories than females. In the Bank vole home ranges in one study were found to be 1.7 acres (0.7ha) for females and 2 acres (0.8ha) for males. Territory size decreases as population density rises. In the Prairie vole home range length drops from 82ft (25m) at low density to 33ft (10m) at high density.

Many species live in areas of little or no direct agricultural importance to man or in areas little changed by human habitation. These species are neither pests nor are they endangered. Nevertheless, species living in temperate areas can be serious agricultural pests and damage pastures, crops, orchards, vineyards and commercial forests. The American pine vole burrows in winter around the base of apple trees and chews on the roots or girdles the stems, resulting in a considerable loss in apple production. The extensive underground burrows of Brandt's vole can become a danger to grazing stock, while muskrats can damage irrigation ditches.

In addition to their status as pests many species harbor vectors of diseases such as plague (some *Microtus* and mole-lemmings) and sylvatic plague (Sagebrush vole). *Microtus* and water voles carry tularemia (an infectious disease which causes fever, chills and inflammation of the lymph glands) in eastern Europe and central Asia. The clearing of forest in these areas has increased the habitat for *Microtus* and also the incidence of tularemia.　　　　　　　　　　　　CJK

Lemming Migrations

Their function in the life cycle of the Norway lemming

According to Scandinavian legend, every few years regimented masses of lemmings descend from birch woods and invade upland pastures where they destroy crops, foul wells and, with their decomposing bodies, infect the air, causing terror-stricken peasants to suffer from giddiness and jaundice. Driven by an irresistible compulsion, not pausing at obstacles—neither man nor beast, river nor ravine—they press on with their suicidal march to the sea. Another story tells of their origins: lemmings are spontaneously generated from foul matter in the clouds and fall to earth during storms and sudden rains. The Eskimos similarly considered them to be not of this world: the name for one North American species means "creatures from outer space."

The life cycle of Norway lemmings is not exceptional. They are active throughout the year. During winter they tunnel and build nests under the snow where they breed and remain safe from predators, except for an occasional attack by an ermine or weasel. The frequency with which they come to the surface depends on the depth and continuity of the snow cover and on the availability of food. With the coming of spring thaw, the lemmings' burrows are in danger of flooding and collapse so the animals are forced to move to higher ground in the alpine zone or lower down in parts of the birch-willow forest. Between May and August they spend much time in the safety of depressions and cavities in the ground or tunnel through the shallow layers of soil and vegetation. At these times of year their enemies include Snowy owls, skuas, ravens and other birds of prey. Well-marked paths and runways often betray the lemmings' presence even when they themselves are out of sight. In the fall, with the freezing of the ground and withering of the sedges, there is a seasonal movement back to sheltered places in the alpine zone. Lemmings are particularly vulnerable at this time: should freezing rain and frost blanket the vegetation with ice before the establishment of snow cover, the

difficulty in gathering food can be fatal.

The mass migrations that have made the Norway lemming famous usually begin in the summer or fall if there has been a period of rapid increase in numbers. In years of low density there may be 1.2–20 lemmings per acre (3–50 per ha) but in peak years 134 per acre (330 per ha) or more. Initially the migrations appear as a gradual movement from densely populated areas in mountain heaths downwards into the willow, birch and conifer forests. The lemmings appear to wander at random. However, as a migration continues groups of animals may be forced to coalesce by local topography. For instance, a large lake may block a migratory movement and thus halt the onward flow, or the lemmings may be caught in a funnel where two rivers meet. In such situations the continuous accumulations of animals become so great that a sort of mass panic ensues and the animals take to wreckless flight—upwards, over rivers, lakes and glaciers and occasionally into the sea.

Although the causes of mass migrations are far from certain, it is widely believed that they are triggered by overcrowding.

▲ **The first known representation of lemmings** appeared in 1555, a woodcut illustration in *Historia de gentibus septentrionalibus* ("History of the Northern Peoples") by the Swedish Catholic priest Olaus Magnus. Not only does the print allude to their origins in the clouds but shows a lemming migration.

▶ **The face of the Norway lemming.** This lemming's normal distribution covers the tundra region of Scandinavia and northwest Russia, in normal years. During "lemming years" the distribution expands considerably.

▶ **A watery grave.** BELOW The main cause of death on lemming migrations is drowning. All such deaths are in a sense accidental. Lemmings can swim well and normally take to water only when they feel they can cross successfully to a far shore. Changes in conditions, eg of the weather, make their destination to be not on this earth.

▼ **Norway lemmings are individualistic,** intolerant creatures. When numbers rise so does aggressive behavior, which is well developed. Here (1) two males box, (2) wrestle and (3) threaten.

1

2

3

Females can breed in their third week and males as early as their first month; reproduction continues year round. Litters consist of five to eight young and may be produced every three or four weeks. In a short period lemmings can produce several generations and increase the overall population rapidly. Short winters without sudden thaws or freezes followed by an early spring and late fall provide favorable conditions for continuous breeding and a rapid increase in population density.

Lemmings are generally intolerant of one another and, apart from brief encounters for mating, lead solitary lives. According to one 19th-century naturalist, Robert Collett (1842–1912): "The enormous multitudes [during peak years] require increased space, and individuals ... cannot, on account of their disposition, bear the unaccustomed proximity of neighbours. Involuntarily the individuals are pressed out to the sides until the edge of the mountain is reached." A want of food is apparently not important in initiating mass migrations since enough food appears to be available even in areas where lemmings are most numerous. It is possible that during a peak year the number of aggressive interactions increases drastically and that this triggers migrations. This seems to be confirmed by reports that up to 80 percent of migrating Norway lemmings are young animals (and thus are likely to have been defeated by larger individuals).

"Lemming years" often correlate with marked increases in the number of other animals living in the same area, for example shrews and voles, capercaillies, and even certain butterflies (whose caterpillars strip entire birch forests of their leaves). But the essential feature of long-distance migrations appears to be a desire to ensure survival. The lemming species of Alaska and northern Canada (the Brown lemming and the Collared lemming) also engage in similar if less spectacular migrations. Although countless thousands of Norway lemmings may perish on their long journeys, the idea that these migrations always end in mass suicide is nonsense.　　　UWH

OLD WORLD RATS AND MICE

Subfamily: Murinae
Four hundred and eight species in 89 genera.
Family: Muridae.
Distribution: Europe, Asia, Africa (excluding
Madagascar), Australia; also found on many
offshore islands.

Habitat: grassland, forest, mountains.

Size: ranges from head-body length 1.7–3.2in
(4.5–8.2cm), tail length 1.1–2.5in
(2.8–6.5cm), weight about 0.2oz (6g) in the
Pygmy mouse to head-body length 19in
(48cm), tail length 8–13in (20–32cm) in
Cuming's slender-tailed cloud rat.

Gestation: in most small species 20–30 days
(longer in species that give birth to precocious
young, eg Spiny mice 36–40 days); not known
for large species.

Longevity: small species live little over 1 year;
Bandicota indica (32oz) 21 days.

THE Old World rats and mice, or Murinae, include 406 species distributed over the major Old World land masses from immediately south of the Arctic Circle to the tips of the southern continents. If exuberant radiation of species, ability to survive, multiply and adapt quickly are criteria of success, then the Old World rats and mice must be regarded as the most successful group of mammals.

They probably originated in Southeast Asia in the late Oligocene or early Miocene (about 25–20 million years ago) from a primitive (cricetine) stock. The earliest fossils (*Progonomys*), in a generally poor fossil representation, are known from the late Miocene (about 8–6 million years ago) of Spain. Old World rats and mice are primarily a tropical group which have sent a few hardy migrant species into temperate Eurasia.

Their success lies in the combination of features they have probably inherited and adapted from a primitive, mouselike "archimurine." This is a hypothetical form, but many features of existing species point to such an ancestor, from which they are little modified. The archimurine would have been small, perhaps about 4in (10cm) long in head and body with a scaly tail of similar length. The appendages would have been of moderate length thereby facilitating subsequent elongation of the hind legs in jumping forms and short robustness in the forelimbs of burrowers. It would have a full complement of five fingers and toes. The sensory structures (ears, eyes, whiskers and

olfactory organs) would have been well developed. Teeth would have consisted of continuously growing, self-sharpening incisors and three elaborately rasped molar teeth in each side of each upper and lower jaw, with powerful jaw muscles for chewing a wide range of foods and preparing material for nests. The archimurine would have had a short gestation period, would have produced several young per litter and therefore would have multiplied quickly. With its small size the archimurine could have occupied a wide variety of microhabitats. Evolution has produced a wide range of adaptations, but only a few, if highly significant, lines of structural change.

Modifications to the tail have produced organs with a wide range of different capabilities. It has become a long balancing organ, with (as in the Australian hopping mice) or without (as in the Wood mouse) a pencil of hairs at its tip. It has become a

Species and genera include:

African creek rat (*Pelomys isseli*), **African forest rat** (*Praomys jacksoni*), **African grass rats** (genus *Arvicanthis*), **African marsh rat** (*Dasymys incomtus*), **African meadow rat** (*Praomys fumatus*), **African soft-furred rats** (genus *Praomys*), **African swamp rats** (genus *Malacomys*), **Australian hopping mice** (genus *Notomys*), **Bushy-tailed cloud rat** (*Crateromys schadenbergi*), **Chestnut rat** (*Niviventer fulvescens*), **Climbing wood-mouse** (*Praomys alleni*), **Cuming's slender-tailed cloud rat** (*Phloeomys cumingi*), **Eastern small-toothed rat** (*Macruromys major*), **Edwards long-footed rat** (*Malacomys edwardsi*), **Four-striped grass mouse** (*Rhabdomys pumilio*), **giant naked-tailed rats** (genus *Uromys*), **Greater tree mouse** (*Chiruromys forbesi*), **harsh-furred rats** (genus *Lophuromys*), **Harvest mouse** (*Micromys minutus*), **Hind's bush rat** (*Aethomys hindei*), **House mouse** (*Mus musculus*), **Larger pygmy mouse** (*Mus triton*), **Lesser bandicoot rat** (*Bandicota bengalensis*), **Lesser ranee mouse** (*Haeromys pusillus*), **Long-footed rat** (*Malacomys longipes*), **Mimic tree rat** (*Xenuromys barbatus*), **New Guinea jumping mouse** (*Lorentzimys nouhuysi*), **Nile rat** (*Arvicanthis niloticus*), **Norway**, **Brown** or

Common rat (*Rattus norvegicus*), **Old World rats** (genus *Rattus*), **Oriental spiny rats** (genus *Maxomys*). **Palm mouse** (*Vandeleuria oleracea*), **Pencil-tailed tree mouse** (*Chiropodomys gliroides*), **Peter's arboreal forest rat** (*Thamnomys rutilans*), **Peter's striped mouse** (*Hybomys univittatus*), **Polynesian rat** (*Rattus exulans*), **Punctated grass-mouse** (*Lemniscomys striatus*), **Pygmy mouse** (*Mus minutoides*), **Roof rat** (*Rattus rattus*), **Rough-tailed giant rat** (*Hyomys goliath*), **Rufous-nosed rat** (*Oenomys hypoxanthus*), **Short-tailed bandicoot rat** (*Nesokia indica*), **small-toothed rats** (genus *Macruromys*), **Smooth-tailed giant rat** (*Mallomys rothschildi*,), **Speckled harsh-furred rat** (*Lophuromys flavopunctatus*), **spiny mice** (genus *Acomys*), **Stick-nest rat** (*Leporillus conditor*), **Striped grass mouse** (*Lemniscomys barbarus*), **Temminck's striped mouse** (*Hybomys trivirgatus*), **Tree rat** (*Thamnomys dolichurus*), **Western small-toothed rat** (*Macruromys elegans*). **Wood mouse** (*Apodemus sylvaticus*).

▲ **The night shift.** Yellow-necked field mice—tree-dwelling nocturnal creatures distributed across much of Europe and Israel—are important for nature as distributors of seeds. This particular mouse is carrying a hazel nut, probably for food. Its rodent incisor teeth enable it both to transport the nut easily and to open it.

grasping organ to help in climbing, as in the Harvest mouse. It has become a sensory organ with numerous tactile hairs at the end furthest from the animal as well as being prehensile (as in the Greater tree mouse). In some it has become a bushy structure, as in the Bushy-tailed cloud rat. In some genera, for example spiny mice, as in some lizards, the tail is readily broken, either in its entirety or in part. Unlike lizards' tails it does not regenerate. In species where the proximal part of the tail is a dark color and the distal part white (for example the Smooth-tailed giant rat) the tail may serve as an organ of communication

Hands and feet show a similar range of adaptation. In climbing forms big toes are often opposable though sometimes relatively small (eg Palm mouse, Lesser ranee mouse). The hands and/or feet can be broadened to produce a firmer grip (eg Pencil-tailed tree mouse, Peter's arboreal

forest rat). In jumping forms the hind legs and feet may be much elongated (eg Australian hopping mice), while in species living in wet, marshy conditions the hind feet can be long and slightly splayed (eg African swamp rats), somewhat reminiscent of the webbed foot of a duck.

The claws are often modified to be short and recurved, for attaching to bark and other rough surfaces (eg Peter's arboreal forest rat) or to be large and strong, in burrowing forms (eg Lesser bandicoot rat). In some of the species with a small opposable digit the claw of this digit becomes small, flattened and nail-like (eg Pencil-tailed tree mouse).

Fur is important for insulation. In some species some hairs of the back are modified into short stiff spines (eg spiny mice) while in others it can be bristly (eg harsh-furred rats), shaggy (eg African marsh rat) or soft and woolly (eg African forest rat), with many intermediates. The function of spines is not known, although it is speculated that they deter predators.

The coat patterns of many species are dominated by medium to dark browns on the back and flanks. Others are designed to conceal an animal. In the Four-striped grass mouse and striped grass mice alternating stripes of dark brown and yellow-brown run along the length of the body so that during the day or evening they can blend with their environments. Mice that live in the deserts of Australia (eg Australian hopping mice) have a sandy brown fur. Many species have a white, gray or light-colored underside. This may be protective, because shadows can make the lighter underside look the same tone as the upper body.

Ears can range from the large, mobile and prominent (as in the Stick-nest rat) to the small and inconspicuous, well covered by surrounding hair (eg African marsh rat). In teeth there is considerable adaptation in the row of molars. In what is presumed to be the primitive condition there are three rows of three cusps on each upper molar tooth. The number of cusps is often much smaller, particularly in the third molar which is often small. The cusps may also coalesce to form transverse ridges. But the typical rounded cusps, although they wear with age, make excellent structures for chewing a wide variety of foods.

The adaptations of teeth have, at the extremes, resulted in the development of robustness and relatively large teeth (in Rusty-nosed rat, Nile rat) and in the reduction of the whole tooth row to a relatively small size, as in the Western small-toothed rat of New Guinea. The food of this rat probably requires little chewing. It possibly consists of soft fruit or small insects.

Murines are found throughout the Old World with considerable variations in the numbers of species in different parts of their range, though in examining their natural distribution the House mouse, Roof rat, Norway rat and Polynesian rat must be discounted as they have been inadvertently introduced in many parts of the world.

The north temperate region is poor in species with, in Europe, countries such as Norway, Great Britain and Poland having respectively as few as 2, 3 and 4 species each. In Africa the density of species is low from the north across the Sahara until the savanna is reached where the richness of species is considerable. Highest densities occur in the tropical rain forest and in adjacent regions of the Congo basin. This can be illustrated by reference to selected locations. The desert around Khartoum, the arid savanna at Bandia, Senegal, the moist savanna in Rwenzori Park, Uganda, and the rainforests of Makokou, Gabon, support 0, 6, 9 and 13 species respectively. Zaire boasts 44 species, and Uganda 36.

Moving to the Orient, species are most numerous south of the Himalayas. India and Sri Lanka have about 35 species, Malaya 22. In the East Indies some islands are remarkably rich: about 41 species in New Guinea, 33 in Sulawesi (Celebes), 32 in the Philippines. Within the Philippines there has been a considerable development of native species with 10 of the 12 genera and 30 species, found only there (ie endemic); only two species of *Rattus* are found elsewhere. A notable feature is the presence of 10 large species having head-body lengths of about 8in (20cm) or more. The largest known murine is found in the Philippines, Cuming's slender-tailed cloud rat, over 16in (40cm) in head-body length. Just slightly shorter are the Pallid slender-tailed rat and the Bushy-tailed cloud rat. This high degree of endemism and the tendency to evolve large species is also found in the other island groups. In New Guinea there are 6 species with head-body lengths of more than 12in (30cm) (included the Smooth-tailed giant rat, the Rough-tailed giant rat, the Eastern small-toothed rat, the giant naked-tailed rats and the Mimic tree rat) and only one small species, the New Guinea jumping mouse, about the size of the House mouse. In Australia there are about 49 species of which approximately 75 percent are to be found in the eastern half of the continent and 55 percent in the west.

It has proved difficult to give an adequate and comprehensive explanation of the evolution and species richness of the murines. There are some pointers to the course evolution may have followed, based on structural affinities and ecological considerations. The murines are a structurally similar group and many of their minor modifications are clearly adaptive, so there are few characters that can be used to distinguish between, in terms of evolution, primitive and advanced conditions. In fact only the row of molar teeth has been used in this way: primitive dentition can be recognized in the presence of a large number of well-formed cusps. Divergences from this condition may represent specialization or advancement. Ecological considerations account for abundance and for the types of habitat preference a species may show. From this analysis two groups of genera have been recognized. The first contains the

▶ ▼ **Old World rats and mice.** (1) A spiny mouse (genus *Acomys*). (2) The Pencil-tailed tree mouse (*Chiropodomys gliroides*). (3) The African marsh rat (*Dasymys incomtus*). (4) The Harsh-furred mouse (*Lophuromys sikapusi*) eating an insect. (5) The Multimammate rat (*Praomys natalensis*). (6) The Four-striped grass mouse (*Rhabdomys pumilio*). (7) The Fawn-colored hopping mouse (*Notomys cervinus*). (8) The Smooth-tailed giant rat (*Mallomys rothschildi*). (a) Tail of the Bushy-tailed cloud rat (*Cratheromys schadenbergi*). (b) Tail of the Greater tree mouse (*Chiruromys forbesi*). (c) Tail of the Harvest mouse (*Micromys minutus*). (d) Tail of a wood mouse (genus *Apodemus*). (e) Hindfoot of the Palm mouse (*Vandeleuria oleracea*); first and fifth digits opposable to provide grip for living in trees. (f) Hindfoot of Peter's arboreal forest rat (*Thamnomys rutilans*); has broad, short digits for providing grip. (g) Paw of the Lesser Bandicoot rat (*Bandicota bengalensis*) showing long, stout claws. (h) Hindfoot of an African swamp rat (genus *Malacomys*) showing long, splayed foot with digits adapted for walking in swampy terrain.

dominant genera (African soft-furred rats, Oriental spiny rats, Old World rats, giant naked-tailed rats, *Mus*, African grass rats and African marsh rats) which have been particularly successful, living in high populations in the best habitats. These are believed to have evolved slowly because they display relatively few changes from the primitive dental condition. The second group contains many of the remaining genera which are less successful, living in marginal habitats and often showing a combination of aberrant, primitive and specialized dental features.

The dominant genera (with the exception of the African marsh rat) contain more species than the peripheral genera and are constantly attempting to extend their range. Considerable numbers of new species have apparently arisen within what is now the range center of a dominant genus (eg soft-furred rats in central Africa and Old World rats in Southeast Asia). The reasons for this await explanation.

It is quite common for two or more species of murine to occur in the same habitat, particularly in the tropics. One of the more interesting and important aspects of studies is to explain the ecological roles assumed by each species in a particular habitat, and then to deduce the patterns of niche occupation and the limits of ecological adaptations by animals with a remarkably uniform basic structure. A particularly favorable habitat and one amenable to this type of study is regenerating tropical forest.

In Mayanja Forest, Uganda, 13 species were found in a recent study from a small area of about 1.5sq mi (4sq km). Certain species were of savanna origin and restricted to grassy rides, ie Rusty-bellied rat and Punctated grass mouse. Of the remaining 11 species all have forest and scrub as their typical habitat with the exception of the two smallest species, the Pygmy mouse and the Larger pygmy mouse, which are also found in grasslands and cultivated areas. Three species, the Tree rat, the Climbing wood-mouse and Peter's arboreal forest rat, seldom, if ever, come to the ground. The small Climbing wood mouse preferred a bushy type of habitat, being frequently found within the first 24in (60cm) off the ground. The two other arboreal species were strong branch runners and were able to exploit the upper as well as the lower levels of trees and bushes. All three species were found alongside a variety of plant species (in the case of the wood mouse, 37 were captured beside 19 different plants, with *Solanum* among the most favored). All species are herbivorous

Rats, Mice and Man

Some Old World rats and mice have a close detrimental association with man through consuming or spoiling his food and crops, damaging his property and carrying disease.

The most important species commensal with man are the Norway or Brown or Common rat, the Roof rat and the House mouse. Now of worldwide distribution, they originated from around the Caspian Sea, India and Turkestan respectively. While the Roof rat and the House mouse have been extending their ranges for many hundreds of years the Norway rat's progress has been appreciably slower, being unknown in the west before the 11th century. The Norway rat is well established in urban and rural situations in temperate regions, and it is the rodent of sewers. In the tropics it is mainly restricted to large cities and ports. The Roof rat is more successful in the tropics where towns and villages are often infested, though it cannot compete with the indigenous species in the field. In the absence of competitors in many Pacific, Atlantic and Caribbean islands, Roof rats are common in agricultural and natural habitats.

With even the solitary House mouse capable of considerable damage in one's home, the scale of mass outbreak damage is difficult to envisage, as when an Australian farmer recorded 28,000 dead mice on his veranda after one night's poisoning, and 70,000 were killed in a wheat yard in an afternoon. In addition to these cosmopolitan commensals there are the more localized Multimammate rat in Africa, the Polynesian rat in Asia and the Lesser bandicoot rat in India.

There are many rodent-borne diseases transmitted either through an intermediate host or directly. The Roof rat, along with other species, hosts the plague bacterium which is transmitted through the flea *Xenopsylla cheopsis*. The lassa fever virus of West Africa is transmitted through urine and feces of the Multimammate rat. Other diseases in which murines are involved include murine typhus, rat-bite fever and leptospirosis.

▲ **Tiny nimble-footed rodents:** the Old World harvest mouse. These are among the smallest of rodents and have a highly developed way of life in tall vegetation, for which they have equipped themselves well. Their legs and feet are highly flexible, their tail able to grasp. They use tall grasses or reeds as if they were a gymnasium, or a forest, building their nests of woven grass high above ground.

▲ **Peripheral rat.** ABOVE LEFT Most of the 63 species of rats belonging to the genus *Rattus* live well away from human habitation, including this bush rat from Australia.

projecting entrances in the shrub layer. They are constructed of grass, on which this species is known to feed.

Of the 11 forest species, Peter's striped mouse, the Speckled harsh-furred rat, the Long-footed rat, the Pygmy mouse, the Larger pygmy mouse and Hind's bush rat are ground dwellers. Of these the striped mouse is a vegetarian, preferring the moister parts of the forest, the Harsh-furred rat an abundant species, predominantly predatory, favoring insects but also prepared to eat other types of flesh. The Long-footed rat is found in the vicinity of streams and swampy conditions; it is nocturnal and includes in its diet insects, slugs and even toads (a specimen in the laboratory constantly attempted to immerse itself in a bowl of water). The two small mice are omnivores and Hind's bush rat is a vegetarian species that inhabits scrub.

A further important feature, which could well account for dietary differences in these species, is the size ranges they occupy. The three mice are in the 0.2–0.9oz (5–25g) range with the Pygmy mouse rather smaller than the other two. The Rufous-nosed rat is in the 2.5–3.2oz (70–90g) range and the Long-footed rat and Hind's bush rat above this. The remaining species, the Tree rat, Peter's arboreal forest rat, African forest rat, Speckled harsh-furred rat and Peter's striped mouse, have weights between 1.2 and 2.1oz (35–60g).

Within the tropical forests there is a high precipitation, with rain falling in all months of the year. This results in continuous flowering, fruiting and herbaceous growth which is reflected in the breeding activity of the rats and mice. In Mayanja Forest the African forest rat and the Speckled harsh-furred rat were the only species obtained in sufficient numbers to permit the monthly examination of reproductive activity. The former bred throughout the year while in the latter the highest frequency of conception coincided with the wetter periods of March to May and October to December.

The foregoing account has attempted to highlight the species richness and adaptability of this group of mammals. In spite of their abundance and ubiquity in the Old World, particularly the tropics, the murines remain a poorly studied group. Exceptions include a few species of economic importance and the Palaearctic wood mouse. Many species are known only from small numbers in museums supported by the briefest information on their biology. There are undoubtedly endless opportunities for research on this fascinating and accessible group of mammals.   MJD

and nocturnal and the two larger species construct elaborately woven nests of vegetation.

Two species are found on both the ground and in the vegetation up to 6.5ft (2m) above it. Of these the African forest rat was abundant and the Rufous-nosed rat much less common. The African forest rat lives in burrows (in which it builds its nest); their entrances are often situated at the bases of trees. It is nocturnal, feeding on a wide range of insect and plant foods. The Rufousnosed rat is both nocturnal and active by day and constructs nests with downwardly

An Animal Weed

Man's relationship with the House mouse

At least since the beginning of recorded history the presence of the common House mouse has compelled man to form a view of its nature and constitution. The first written reference to the raising and protection of mice by man amounts to evidence of mouse worship in Pontis (Asia Minor) 1,400 years before the birth of Christ. Homer mentions Apollo Smintheus, a god of mice, about 1200 BC, and his worship was still popular at the time of Alexander the Great 900 years later. Mice were worshiped by the Teucrans of Crete, who attributed their victory over the Pontians to a God who caused mice to gnaw the leather straps on the shields of their enemies. A temple at Tenedos on the entrance to the Dardanelles (whose foundations still remain) was built in which mice were maintained at public expense. The mouse cult spread to other cities in Greece and continued as a local form of worship until the Turkish conquest in 1543. Pliny wrote "They (white mice) are not without certain natural properties with regard to the sympathy between them and the planets in their ascent ... Sooth-sayers have observed that it is a sign of prosperity if there be a store of white ones bred." Hippocrates recorded that he did not need to use mouse blood as a cure for warts in the same way as his colleagues because he had a magic stone with lumps on it which had proved to be an efficient remedy.

The other main center of ancient mouse culture was in the Far East. In China, albino mice were used as auguries, and Chinese government records show that 30 albino mice were caught in the wild between AD 307 and 1641. A word for "spotted mouse" appears in the first Chinese dictionary, written in 1100 BC. Waltzing mice have been known since at least 80 BC (waltzing or dancing in mice is produced by a defect in the inner ear affecting balance, which is usually inherited). Clearly there has been a mouse fancy in China (and Japan) for many centuries which valued new and unusual forms. The Japanese had in their fancy such traits as albinism, non-agouti, chocolate, waltzing, dominant and recessive spotting, and other color variants still known to mouse breeders. Some of these varieties were brought to Europe in the mid-19th century by British traders.

In western Europe as early as 1664 the English scientist Robert Hooke used a mouse to study the effects of increased air pressure. William Harvey (1578–1657) used mice in his anatomical studies. The English chemist Joseph Priestly gives a delightful account in his *Experiments and Observations* (1775) of

his experiments with mice, including how he trapped and maintained them. Half a century after this, a Genevan pharmacist named Louis Coladon bred large numbers of white and gray mice and obtained segregations in agreement with Mendelian expectation 36 years before Mendel published his work on peas. One of the earliest demonstrations of genetical segregation in animals after the rediscovery of Mendel's papers was made by L. Cuenot working in Paris with mice (1902).

The modern history of laboratory mouse breeding began in 1907 when an undergraduate at Harvard University, C.C. Little, began to study the inheritance of coat color under the supervision of W.E. Castle. Two years later, Little obtained a pair of "fancy" mice carrying alleles (variations of genes) for the recessively inherited traits dilution (*d*), brown (*b*), and non-agouti (*a*), and inbred their descendants brother to sister, selecting for vigorous animals. The recessive trait only appears when an animal inherits the *same* allele from both its parents (each parent contributes one allele of the pair in the offspring). If the trait appears when an animal has only one allele, it is said to be dominantly inherited. Inbreeding has the

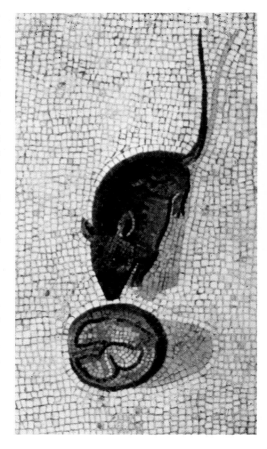

▲ **The variable House mouse.** Genetic diversity is exemplified in this nest of laboratory mice by the color variations in the young.

◀ **A delight in the mouse** is shown in this detail from a Roman mosaic.

▼ **Man's constant companions.** The wide distribution of the House mouse may be due to its characteristics or could be a historical development, resulting from its occurrences with Neolithic man (around 10,000 years ago) in the first cereal-growing sites in the Near East and subsequently spread with increasing cereal production.

effect that animals will share ancestors, and therefore the likelihood that they will get the same allele for a particular characteristic from both parents will be much greater than if they were unrelated. A strain which has been inbred will breed true for inherited characteristics. The first truly inbred strain (exclusively brother-sister, mated for over 20 generations) was Little's DBA strain (called after the three mutant genes it carried). Because all the animals in an inbred strain carry the same genes, the effects of different treatments (drugs, physical environment, etc) can be compared without any confusion being produced by variation due to genetic differences.

There are now many other inbred strains, most of them descended from mice caught in the northeastern USA. Laboratory workers have recently become increasingly interested in wild mice, because they carry many inherited variants not found in established inbred strains, and allow us to find out more about development and immunology.

The House mouse is an animal weed—quick to exploit opportunity and able to withstand local adversity and extinction without the species as a whole being harmed. This means that it must be able to breed rapidly, tolerate a wide range of conditions, and adjust quickly to changes in environment. These traits are responsible for success in all but a few parts of the world where it is excluded by extreme environments (eg polar regions) or competition from other small mammals (eg Central Africa). Whether all *Mus* species are equally adaptable is unknown.

If mice adjusted to their environments exclusively genetically it would be easy to summarize the situation. However, mouse populations differ considerably in their genetical composition (genomes), and different genomes respond differently to similar environmental pressures. For example, the tail of the mouse seems to be a heat-regulating organ; generally, relative tail length is greater in mice reared in hot than in cold conditions but the genome is the same. Tail length decreases from about 95 percent of head and body length in southern England to less than 80 percent in Orkney—but mice from the Shetland and Faroe groups north of Orkney have tails as long as southern English animals.

House mice are among the more genetically variable mammals, with the most variable population known in Hawaii.   RJB

OTHER OLD WORLD RATS AND MICE

Ninety-three species in 45 genera belonging to 10 subfamilies
Distribution: Balkans, S USSR, Africa, Madagascar, S India, China, New Guinea, Australia.

Blind mole-rats   **African pouched rats**   **Crested rat**

African climbing mice   **Root and Bamboo rats**   **Madagascan rats**

African swamp rats   **Oriental dormice**   **Zokors**   **Australasian water rats**

THROUGHOUT the Old World there are small groups of rats and mice that cannot be included in the three major subfamilies of the muridae. Here they are grouped in 10 small subfamilies. Relationships amongst these are not clear and they are placed in a geographical sequence for convenience. Some of these rodents are superficially very similar to members of the major groups; eg the zokors can be considered as specialized voles. Two groups, the blind mole-rats and the bamboo rats, are more distinctive and are sometimes treated as separate families.

Of all the subterranean rodents the **blind mole-rats** (subfamily Spalacinae) show the most extreme adaptations to life underground and they are often treated as a separate family, Spalacidae. Their eyes are completely and permanently hidden under their skin and there are no detectable external ears or tail. The incisor teeth protrude so far that they are permanently outside the mouth and can be used for digging without the mouth having to be opened. A unique feature is the horizontal line of short, very stiff (presumably touch-sensitive) hairs on each side of the head. Most blind mole-rats are about 5–9in (13–25cm) long but in one species, the Giant mole-rat of southern Russia, they can reach 14in (35cm).

Blind mole-rats are found in dry but not desert habitats from the Balkans and southern Russia round the eastern Mediterranean as far west as Libya. Apart from being entirely vegetarian they live very much like the true moles (which are predators belonging to the order Insectivora). Each animal makes its own system of tunnels, which may reach as much as 1,150ft (350m) in length, throwing up heaps of soil. They feed especially on fleshy roots—bulbs and tubers —but also on whole plants. Although originally animals of the steppes they have adapted well to cultivation and are a considerable pest in crops of roots, grain and fruit.

Blind mole-rats breed in spring, with usually two or three in a litter which disperse away from the mother's tunnel system as soon as they are weaned, at about three weeks. There is sometimes a second litter later in the year.

As in many other burrowing mammals their limited movement has led to the evolution of many local forms which make the individual species very difficult to define but provides a bonanza for the study of genetics and the processes of evolution and species formation. The number of species that should be recognized and how they are classified are still very uncertain. Eight species are recognized here but they have been reduced to three elsewhere.

The five species of **African pouched rats** (subfamily Cricetomyinae) resemble hamsters in having a storage pouch opening from the inside of each cheek. The two short-tailed pouched rats resemble hamsters in general appearance but the other three species are rat-like, with long tail and large ears.

The giant pouched rats are among the largest of the murid rodents, reaching 16in (40cm) in head-body length. They are common throughout Africa south of the Sahara and in some areas are hunted for food. They feed on a very wide variety of items, including insects and snails as well as seeds and fruit. In addition to carrying food to underground storage chambers, the cheek pouches can be inflated with air as a threat display. The gestation period is about six weeks and litter size usually two or four.

The three large species are associated with peculiar blind, wingless earwigs of the genus *Hemimerus* which occur in their fur and in their nests where they probably share the rat's food.

African swamp rats (or vlei rats) (subfamily Otomyinae) are found throughout much of

4

► ▼ **Representatives of minor subfamilies** of rats and mice. (1) The Spiny dormouse (*Platacanthomys lasiurus*). (2) *Macrotarsomys ingens* (one of the Madagascan rats). (3) The Crested rat (*Lophiomys imhausii*) with mane erect, showing a glandular patch. (4) The East Siberian Zokor (*Myospalax aspalax*), kicking back excavated soil with its hindfeet. (5) The Ehrenberg's mole-rat (*Nannospalax ehrenbergi*) showing the broad nose used for ramming soil and tactile hairs on the face. Note the absence of external eyes.

Africa south of the Sahara. Externally they can hardly be distinguished from some of the voles of the northern temperate region; like the voles of the genera *Microtus* and *Arvicola* they are medium-sized ground-living rodents with short ears and legs, small eyes, a short blunt muzzle and shaggy fur. The tail is shorter than the head and body but not as short as in most voles. The color, as in many voles, is usually a grizzled yellow brown. The various species are very similar and have not yet been adequately defined.

Like many voles most swamp rats live among thick grass, especially in wet areas, and they are particularly abundant in the alpine zone of many African mountains such as Mount Kenya and Kilimanjaro. Where ground vegetation is thick they are active by day as well as by night, making runways at ground level among the grass tussocks. They are adapted to feeding on grass and tough stems, having high-crowned molar teeth with multiple transverse ridges of hard enamel, although they are not ever-growing as in most voles.

Swamp rats have small litters, usually of one or two quite well-developed young.

The two species of whistling rats are very similar to the species of *Otomys* but they are lighter in color and live in dry country with sparse grass and low shrubs. They are sociable animals, active by day, and when they are alarmed outside their burrows they whistle loudly and stamp their feet before disappearing underground.

The **Crested rat** has so many peculiarities that it is placed in a subfamily of its own (Lophiomyinae) and it is not at all clear what are its nearest relatives. It is a large, dumpy, shaggy rodent with a bushy tail and tracts of long hair along each side of the back which can be erected. These are associated with specialized scent glands in the skin and the individual hairs of the crests have a unique lattice-like structure which probably serves to hold and disseminate the scent. These hair tracts can be suddenly parted to expose the bold striped pattern as well as the scent glands.

The skull is also unique in having a peculiar granular texture and in having the cavities occupied by the principal, temporal, chewing muscles roofed over by bone—a feature not found in any other rodent.

Crested rats are nocturnal and little is known of their way of life. They spend the day in burrow, rock crevice or hollow tree. They are competent climbers and feed on a variety of vegetable material. The stomach is unique among rodents in being divided into a number of complex chambers similar to those found in ruminant ungulates such as cattle and deer.

The majority of **African climbing mice** (subfamily Dendromurinae) are small agile mice with long tails and slender feet, adapted to climbing among trees, shrubs and long grass. Though they are confined to Africa south of the Sahara some of them closely resemble mice in other regions that show similar adaptations, such as the Eurasian harvest mouse (subfamily Murinae) and the North American harvest mice (subfamily Hesperomyinae). They are separated from these mainly by a unique pattern of cusps on the molar teeth and it is on the basis of this feature that some superficially very different rodents have been associated with them in the subfamily Dendromurinae.

Typical dendromurines, eg those of the genus *Dendromus*, are nocturnal and feed on grass seeds but are also considerable predators on small insects such as beetles and even young birds and lizards. Some species in other genera are suspected of being more completely insectivorous. In the genus *Dendromus* some species make compact globular nests of grass above ground, eg in bushes; others nest underground. Breeding is seasonal, with usually three to six, naked, blind nestlings in a litter. Among the other genera the most unusual are the fat mice. They make extensive burrows and during the dry season spend long periods underground in a state of torpor, after developing thick deposits of fat. Even during their active season fat mice become torpid, with reduced body temperature, during the day.

Many species of this subfamily are poorly

▲ **Apparatus for life underground.** The East African root rat rarely comes above ground—only for the occasional forage. It digs the underground burrows that provide its environment using its powerful incisor teeth.

▶ **Breeding mound of a blind mole-rat** (*Nannospalax ehrenbergi*). Each animal makes its own system of tunnels which may be as much as 1,150ft (350m) in length, throwing up heaps of soil. These mole-rats feed especially on fleshy roots—bulbs and tubers—but also on whole plants by pulling them down into their tunnels by the roots. They store food: their underground food storage chambers have been known to hold as much as 31lb (14kg) of assorted vegetables. The breeding nest is placed centrally within the mound.

The 10 subfamilies of Other Old World rats and mice

Blind mole-rats
Subfamily Spalacinae
8 species in 2 genera
Balkans, S Russia, E Mediterranean E N Africa, in dry (but not desert) habitats. Genera: *Spalax*, 5 species; *Nannospalax*, 3 species.

African pouched rats
Subfamily Cricetomyinae
5 species in 3 genera
Africa S of the Sahara, in savanna, dry woodland, tropical forest. Genera: **giant pouched rats** (*Cricetomys*), 2 species; **Lesser pouched rat** (*Beamys hindei*); **short-tailed pouched rats** (*Saccostomys*), 2 species.

African swamp rats
Subfamily Otomyinae
13 species in 2 genera
Africa S of the Sahara. Genera: **swamp and vlei rats** (*Otomys*), 11 species; **whistling rats** (*Parotomys*). 2 species.

Crested rat
Subfamily Lophiomyinae
1 species, *Lophiomys imhausi*, Kenya, Somalia, Ethiopia, E Sudan, in mountain forests between 4,000 and 8,900ft.

African climbing mice
Subfamily Dendromurinae
21 species in 10 genera
Africa S of the Sahara, in savanna. Genera: **climbing mice** (*Dendromus*), 6 species; **Large Ethiopian climbing mouse** (*Megadendromus nikolausi*); **Dollman's tree mouse** (*Prionomys batesi*); **Link rat** (*Deomys ferrugineus*); *Dendroprionomys rousseloti*; *Leimacomys buettneri*; *Malacothrix typica*; **fat mice** (*Steatomys*), 6 species; **rock mice** (*Petromyscus*), 2 species; **Delany's swamp mouse** (*Delanymys brooksi*).

Root and bamboo rats
Subfamily Rhizomyinae
6 species in 3 genera
E Africa and SE Asia, in grassland and savanna and in forests respectively. Genera: **bamboo rats** (*Rhyzomys*), 3 species; **Bay bamboo rat** (*Cannomys badius*); **root rats** (*Tachyoryctes*), 2 species.

Madagascan rats
Subfamily Nesomyinae
11 species in 8 genera
Madagascar with 1 species in S Africa, in forest, scrub, grassland and marsh. Genera: *Macrotarsomys*, 2 species; *Nesomys rufus*; *Brachytarsomys albicauda*; *Eliurus*, 2 species; *Gymnuromys roberti*; *Hypogeomys antimena*; *Brachyuromys betsileoensis*; **White-tailed rat** (*Mystromys albicaudatus*).

Oriental dormice
Subfamily Platacanthomyinae
2 species in 2 genera
Spiny dormouse (*Platacanthomys lasiurus*), S India, in forested hills between 1,650 and 3,300ft. **Chinese dormouse** (*Typhlomys cinereus*), S China and N Vietnam, in forest.

Zokors or Central Asiatic mole rats
Subfamily Myospalacinae
6 species in 1 genus (*Myospalax*)
China and Altai Mountains, underground.

Australian water rats
Subfamily Hydromyinae
20 species in 13 genera
Australia, New Guinea, and Philippines, in forest, rivers, marshes. Genera: *Hydromys*, 4 species including **Australian water rat** (*H. chrysogaster*); **Coarse-haired water rat** (*Parahydromys asper*); **Earless water rat** (*Crossomys moncktoni*); **False swamp rat** (*Xeromys myoides*); **shrew-rats** (*Rhynchomys*), 2 species; **striped rats** (*Chrotomys*), 2 species; **Luzon shrew-rat** (*Celaenomys siliceus*); **shrew-mice** (*Pseudohydromys*), 2 species; **Groove-toothed shrew-mouse** (*Microhydromys richardsoni*); **One-toothed shrew-mouse** (*Mayermys ellermani*); **Short-tailed shrew-mouse** (*Neohydromys fuscus*); **Long-footed hydromyine** (*Leptomys elegans*); *Paraleptomys*, 2 species.

known. Several distinctive new species have been discovered only during the last 20 years and it is likely that others remain to be found, especially arboreal forest species.

The genera *Petromyscus* and *Delanymys* have sometimes been separated from the others in a subfamily Petromyscinae.

Root rats (subfamily Rhizomyinae) are large rats adapted for burrowing and show many of the characteristics found in other burrowing rodents—short extremities, small eyes, large protruding incisor teeth and powerful neck muscles reflected in a broad angular skull. The **bamboo rats**, found in forests of Southeast Asia, show all these features in less extreme form than do the root rats of East Africa. They make extensive burrows in which they spend the day but they emerge at night and do at least some of their feeding above ground. The principal diet consists of the roots of bamboos and other plants, but above-ground shoots are also eaten. In spite of their size breeding is similar to the normal murid pattern with a short gestation of three to four weeks and three to five naked, blind young in a litter, though weaning is protracted and the young may stay with the mother for several months.

The **African root rats** are more subterranean than the bamboo rats but less so than African mole-rats (family Bathyergidae) or the blind mole-rats (subfamily Spalacinae). They make prominent "mole hills" in open country. Roots and tubers are stored underground. As in most mole-like animals each individual occupies its own system of tunnels. The gestation period is unusually long for a murid—between six and seven weeks.

It has long been debated whether the **Madagascan rats**, the ten or so indigenous rodents of the island of Madagascar, form a single, closely interrelated group, implying that they have evolved from the same colonizing species, or whether there have been multiple colonizations such that some of the present species are more closely related to mainland African rodents than to their fellows on Madagascar. The balance of evidence seems to favor the first alternative hence their inclusion here in a single subfamily, Nesomyinae, but the matter is by no means settled.

The inclusion of the South African white-tailed rat is also debatable, the implication being that it is the sole survivor on the African mainland of the stock that colonized Madagascar. The problem arises from the diversity of the Madagascan species, especially in dentition, coupled with the fact

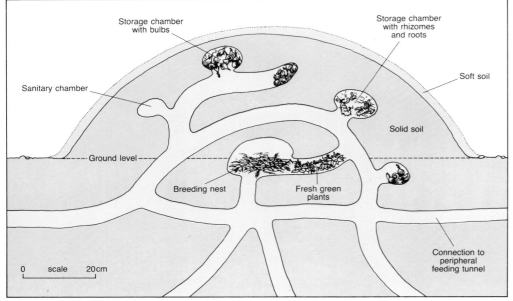

Storage chamber with bulbs

Storage chamber with rhizomes and roots

Sanitary chamber

Soft soil

Solid soil

Ground level

Breeding nest

Fresh green plants

Connection to peripheral feeding tunnel

0   scale   20cm

that none of them match very closely any of the non-Madagascan groups of murid rodents.

The group includes some species that are typical, small, agile mice with long tail, long slender hind feet and large eyes and ears (eg *Macrotarsomys bastardi* and *Eliurus minor*). *Nesomys rufus* is typically rat-like in its size and proportions while *Hypogeomys antimena* is rabbit-sized and makes deep burrows, although it forages for food on the surface.

The two species of *Brachyuromys* are remarkably vole-like in form, dentition and ecology. They live in wet grassland or marshes and are apparently adapted to feeding on grass. Externally they can only with difficulty be distinguished from Eurasian water voles.

The two species of **Oriental dormice** (subfamily Platacanthomyinae) have been considered to be closely related to the true dormice (family Gliridae) which they resemble externally and in the similar pattern of transverse ridges on the molar teeth, although there are only three molars on each row, not preceded by a premolar as in the true dormice. More recently opinion has swung towards treating them as aberrant members of the family Muridae (in its widest sense as used here). Whatever their affinities, they are very distinctive, arboreal mice with no very close relatives, and very little is known of their way of life. The Spiny dormouse is mainly a seed-eater and is a pest of pepper crops in numerous parts of southern India.

Zokors (subfamily Myospalacinae) are burrowing, vole-like rodents found in the steppes and open woodlands in much of China and west as far as the Altai Mountains. Although they live almost entirely underground they are less extremely adapted than the blind mole-rats. Both eyes and external ears are clearly visible, though tiny, and the tail is also distinct. Digging is done mainly by the very large claws of the front feet rather than the teeth. The front claws are so enlarged that when the animal moves it folds the fingers backwards and walks on the knuckles. The coat is a rather uniform grayish brown in most species and there is often a white streak on the forehead.

Like the blind mole-rats, zokors feed on roots, rhizomes and bulbs but they occasionally emerge from their tunnels to collect food such as seeds from the surface. Massive underground stores of food are accumulated, enabling the animals to remain active all winter.

Breeding takes place in spring when one litter of up to six young is produced. Their

▼ ◄ **Representatives of minor subfamilies** of Old World rats and mice. (**1**) Brandt's climbing mouse (*Dendromus mesomelas*). (**2**) A Savanna giant pouched rat (*Cricetomys gambianus*) with both pouches full of food. (**3**) A Vlei rat (*Otomys irroratus*) sitting in a grass runway. (**4**) An East African mole-rat (*Tachyoryctes splendens*) burrowing with its incisors. (**5**) An Australian water rat (*Hydromys chrysogaster*) diving.

social organization is little known but the young appear to stay with the mother for a considerable time.

Australian water rats and their allies (subfamily Hydromyinae) are found mostly in New Guinea with a few species in Australia and in the Philippines. Although generally rat-like in superficial appearance the rodents in this group are diverse and it is not at all certain that they form a natural group more closely related to each other than to other murid rodents. The characteristic that appears to unite them is the simplification of the molar teeth, which lack the strong cusps or ridges found in most of the mouse-like rodents, and in some species the molars are also reduced in number, the extreme being seen in the One-toothed shrew-mouse which has only one small, simple molar in each row.

This adaptation is most likely to be related to a diet of fruit or soft-bodied invertebrates, but little information is available on the diet of most species. It is likely that all the members of this group have been derived from rats of the subfamily Murinae, mainly by the simplification of the molar teeth. However, it is quite possible that this could have happened more than once, so that, for example, the shrew-rats of the Philippines and the water rats of Australia and New Guinea may have arisen independently from separate groups of murine rats at different times.

Many of the species show few other peculiarities but two more specialized groups can be recognized. These are the water rats and the shrew-rats. The water rats, of which the best known, and largest, representative is the Australian water rat (weighing 23–44oz (650–1,250g), have broad, webbed hind feet. The Australian water rat is a common and well-known animal, often seen by day (and often mistaken for a platypus) as it hunts underwater for prey such as frogs, fish, mollusks, insects, and crabs. They have been seen to bring mussels out of the water and leave them on a rock until the heat makes them open and easier to extract from the shell. These water rats breed in spring and summer, producing two or even three litters of usually four or five young after a gestation of about 35 days.

The most specialized of the water rats is the Earless water rat, which not only lacks external ears but has longitudinal fringes of long white hairs on the tail, forming a pattern that is remarkably similar to that found in the completely unrelated Elegant water shrew, *Nectogale elegans*, of the Himalayas.

Although not structurally specialized the little-known False water rat has the most specialized and unusual habitat, for it lives in the mangrove swamps of northern Australia. It climbs among the mangroves and does not appear to spend much time swimming.

The shrew-rats of the Philippines have long snouts with slender, protruding incisor teeth like delicate forceps, presumably adapted, as in the true shrews, for capturing insects and other invertebrates. However, the remaining teeth are small and flat-crowned, quite unlike the sharp-cusped batteries found in the true shrews. The two species of the genus *Rynchomys* show this adaptation in extreme form (and have sometimes been placed in a separate subfamily, the Rynchomyinae).

The two species of striped rats, also in the Philippines, are unique in the group in having a bold pattern consisting of a central bright buff stripe along the back, flanked on each side by a black stripe, producing a simplified version of the pattern seen in some chipmunks.

Of the remaining species the five shrew-mice of New Guinea are small smoky-gray mice living in mountain forest. They have more normal incisors than the shrew-rats but have very rudimentary molars. The three species of *Leptomys* and *Paraleptomys* are rat-sized and probably arboreal.   GBC

HAMSTERS

Subfamily: Cricetinae
Twenty-four species in 5 genera.
Family: Muridae.
Distribution: Europe, Middle East, Russia, China.

Habitat: arid or semiarid areas varying from rocky mountain slopes and steppes to cultivated fields.

Size: ranges from head-body length 2–4in (5.3–10.2cm), tail length 0.3–0.4in (0.7–1.1cm), weight 1.8oz (50g) in the Dzungarian hamster to head-body length 7.9–11in (20–28cm), weight (32oz, 900g) (no tail) in the Common or Black-bellied hamster.

Gestation: ranges from 15 days in the Golden hamster to 37 days in the White-tailed rat.

Longevity: 2–3 years.

Mouse-like hamsters
Genus Calomyscus
Iran, Afghanistan, S Russia, Pakistan. Five species including: **Mouse-like hamster** (*C. bailwardi*).

Rat-like hamsters
Genus Cricetulus
SE Europe, Asia Minor, N Asia. Eleven species including: **Korean gray rat** (*C. triton*).

Common hamster or Black-bellied hamster
Cricetus cricetus.
C Europe, Russia.

Golden hamsters
Genus Mesocricetus
E Europe, Middle East. Four species including: **Golden hamster** (*M. auratus*).

Dwarf hamsters
Genus Phodopus
Siberia, Mongolia, N China. Three species including: **Dzungarian hamster** (*P. sungorus*).

Until the 1930s the Golden hamster was known only from one specimen found in 1839. However, in 1930 a female with 12 young was collected in Syria and taken to Israel. There the littermates bred and some descendants were taken to England in 1931 and to the USA in 1938 where they proliferated. Today the Golden hamster is one of the most familiar pets and laboratory animals in the West. The other hamster species are less well known, though the Common hamster has been familiar for many years.

Most hamsters have small, compact, rounded bodies with short legs, thick fur and large ears, and prominent dark eyes, long whiskers and sharp claws. Most have cheek pouches which consist of loose folds of skin starting from between the prominent incisors and premolars and extending along the outside of the lower jaw. When hamsters forage they can push food into the pouches which then expand, enabling them to carry large quantities of food to the underground storage chamber—a very useful adaptation for animals that live in a habitat where food may occur irregularly but in great abundance. The paws of the front legs are modified hands, giving great dexterity to the manipulation of food. Hamsters also use a characteristic forward squeezing movement of the paws as a means of emptying their cheek pouches of food. Common hamsters are reputed to inflate their cheek pouches with air when crossing streams, presumably to create extra buoyancy.

Hamsters are mainly herbivorous. The Common hamster may hunt insects, lizards, frogs, mice, young birds and even snakes, but such prey contributes only a small amount to the diet. Normally hamsters eat seeds, shoots and root vegetables, including wheat, barley, millet, soybeans, peas, potatoes, carrots, beets as well as leaves and flowers. Small items (such as millet seeds) are carried to the hamster's burrow in its pouches, larger items (such as potatoes) in

its incisors. Food is either stored for the winter, eaten on returning to underground quarters, or, in undisturbed conditions, eaten above ground. One Korean gray rat managed to carry 42 soybeans in its pouches. The record for storage in a burrow probably goes to the Common hamster: chambers of this species have been found to contain as much as 198lb (90kg) of plant material collected by one hamster alone. Hamsters spend the winter in hibernation in their burrows, only waking on warmer days to eat food from their stores.

As children's pets, hamsters have a reputation for gentleness and docility. In the wild, however, they are solitary and exceptionally aggressive towards members of their own species. These characteristics may result from intense competition for patchy but locally abundant food resources, but may also serve to disperse population throughout a particular area or habitat. Large species, such as the Common hamster and the Korean gray rat, also behave aggressively towards other species, and have been known to attack dogs or even people when threatened. To defend itself from attack by predators the Korean gray rat may throw itself on its back and utter piercing screams.

Species studied in the laboratory have been shown to have acute hearing. They communicate with ultrasounds (high frequency sounds) as well as with squeaks audible to the human ear. Ultrasounds appear to be most important between males and females during mating and perhaps synchronize behavior. Their sense of smell is also acute. It has recently been shown that the Golden hamster can recognize individuals, probably from flank gland secretions, and that males can detect stages of a female's estrous cycle by odors, and even recognize a receptive female from the odors of her vaginal secretions.

Most hamsters have an impressive capacity for reproduction and become sexually mature soon after weaning (or even during it). Female Common hamsters become receptive to males at 43 days and can give birth at 59 days. Golden hamsters have slightly slower development and become sexually mature between 56 and 70 days. In the wild they probably breed only once, or occasionally twice, per year during the spring and summer months but in captivity they can breed year round. Their courtship is simple and brief, as befits animals that are in general solitary creatures and meet only to copulate. Odors and restrained movement suffice to indicate that the partners are

▲ **The Common hamster** was once widespread across Central Europe and the Soviet Union. Following the introduction of modern agricultural methods, however, its numbers—at least in Europe—have declined dramatically. Its front paws are extremely dextrous, making it easy to place food in its cheek pouches.

◄ **The Dzungarian hamster,** ABOVE, little known to natural historians, lives in Siberia, Mongolia and northern China. Although only a mere 2 in (5cm) or so in length it displays all the adaptations of its larger relatives.

ready and willing to mate. Immature animals or females not in heat will either attack or be attacked by other individuals. As soon as a pair have copulated they separate and may never meet again. The female builds a nest for the young in her burrow from grass, wool and feathers and gives birth after about 16–20 days in the Common hamster. The young are born hairless and blind and are cared for by the female alone. During this time she may live off her food store in another section of the burrow. The young are weaned at about three weeks of age in the Golden hamster. In

the slowest developing species, the Mouse-like hamster, adult coloration and size may not be reached until six months old.

Hamsters are considered serious pests to agriculture in some areas. In some countries dogs are trained to kill them. Chinese peasants sometimes catch the large rat-like hamsters and eat them, and the Common hamster is trapped for its skin. Despite these pressures, none of the hamsters appears to be endangered at the moment, perhaps because most live in regions inhospitable to man and have such a high reproduction rate.   JF

GERBILS

Subfamily: Gerbillinae
Eighty-one species in 15 genera.
Family: Muridae.
Distribution: Africa, Asian steppes from Turkey and SW Russia in the west to N China in the east.

Habitat: desert, savanna, steppe, rocks, cultivated land.

Size: ranges from head-body length 2.4–2.9in (6.2–7.5cm), tail length 2.8–3.7in (7.2–9.5cm), weight 0.3–0.4oz (8–11g) in Henley's (pygmy) gerbil to head-body length 5.9–7.9in (15–20cm), tail length 6.2–8.7in (16–22cm), weight 4–6.7oz (115–190g) in the Antelope rat or Indian red-footed gerbil.

Gestation: 21–28 days.

Longevity: usually 1–2 years.

To most people a gerbil is an attractive pet rodent with large dark eyes and a furry tail. The animal they have in mind, however, is the Mongolian gerbil, just one of the many species of gerbils, jirds and sandrats that belong to the largest group of rodents in Africa and Asia adapted to arid conditions.

The different genera of gerbils are very distinctive; most are either mouse-like or rat-like. Within genera, however, there are numerous ways in which gerbils differ from one another—in size, color, length of tail, the fashioning of the tail tuft, the color of nails. Variations can even be found within well-defined species. Given such complexities it is impossible to be certain about how many species exist and sometimes about the genus to which a species might belong.

Most gerbils live in arid climates in arid habitats, to both of which they have adapted in interesting ways. For any animal to survive and live it must not lose more water than it normally takes in. (Water loss usually occurs by evaporation from the skin, in air exhaled from the lungs, and by urination and defecation.) The predicament of the gerbil is that it has a large body-surface compared with its volume, and that it has to find ways of obtaining water and minimizing loss.

The 15 genera of gerbils

Ammodillus imbellis
N Kenya, Somalia, E Ethiopia inhabiting steppe and desert.

Brachiones przewalskii (**Przewalski's gerbil** or **jird**)
N China and Mongolian Republic inhabiting desert and steppe.

Desmodilliscus braueri (**Short-eared rat**)
Senegal E to S Sudan inhabiting desert, savanna and wooded grassland.

Desmodillus auricularis (**Namaqua gerbil** or **Cape short-toed gerbil**)
S Africa inhabiting desert, savanna, steppe.

Dipodillus, 3 species
W Sahara and African countries bordering the Mediterranean inhabiting desert and semidesert. Species include **Simon's dipodil** (*D. simoni*).

Gerbillurus, 4 species
S Africa inhabiting savanna and desert. Species include **Brush-tailed gerbil** (*G. vallinus*).

Gerbillus (**smaller gerbils**) 34 species
N Africa, Middle East, Iran, Afghanistan, W India, inhabiting desert, semidesert and coastal areas. Species include **Hairy-footed gerbil** (*G. latastei*), **Pallid gerbil** (*G. perpallidus*), **Wagner's gerbil** (*G. dasyurus*).

Meriones (**jirds**), 14 species
N Africa, Turkey, Middle East, Iran, Afghanistan, NW India, Mongolia, N China inhabiting savanna, desert and steppe. Species include **Indian desert jird**
(*M. hurrianae*), **King jird** (*M. rex*), **Mongolian gerbil** (*M. unguiculatus*).

Microdillus peeli
N Kenya, Somalia, E Ethiopia inhabiting desert and steppe.

Pachyuromys duprasi (**Fat-tailed gerbil**)
N African countries bordering on the Sahara, inhabiting desert and semidesert.

Psammomys, 2 species
N African countries bordering on the Sahara inhabiting desert and semidesert. Species include **Fat sand rat** or **Fat jird** (*P. obesus*).

Rhombomys opimus (**Great gerbil**)
Afghanistan, SW USSR, Mongolia and N China inhabiting steppe and desert.

Sekeetamys calurus (**Bushy-tailed gerbil**)
E Egypt, S Israel, Jordan, Saudi Arabia inhabiting desert.

Tatera (**larger gerbils**), 9 species
S and E Africa, SW Asia, inhabiting savanna, steppe and desert.

Taterillus, 7 species
Senegal E to S Sudan and S to Tanzania inhabiting desert, savanna and wooded grassland. Species include: **Emin's gerbil** (*T. emini*), **Harrington's gerbil** (*T. harringtoni*).

The genera are grouped as follows:
Gerbillus group: *Ammodillus, Dipodillus, Gerbillus, Microdillus, Pachyuromys, Sekeetamys*
Meriones group: *Brachiones, Meriones, Psammomys, Rhombomys*
Tatera group: *Desmodilliscus, Desmodillus, Gerbillurus, Tatera, Taterillus*

The gerbil therefore cannot afford to sweat and indeed cannot survive temperatures of about 113°F (45°C) for more than about two hours. During the day it lives below the surface, at a depth of about 20in (50cm), where the temperature is constant during both day and night, at about 68–77°F (20–25°C). Often the burrow entrance is blocked during the day. Most species are therefore nocturnal: the only gerbils that live on the surface in daytime

▲ **The digestive system of the female gerbil** has to extract sufficient liquid from her food to meet the requirements of herself and her young. These are born in a pre-mature condition, and may have to be suckled for about three weeks. The litter of this Mongolian gerbil is of average size.

are the ''northern'' species, for example the Great gerbil or the Mongolian jird, though some jirds that live further south emerge during the day in winter.

Often in the dry world of the gerbil the only food available is dry seeds. The animal's nocturnal activity enables it to make the most of these. By the time it emerges it finds the seeds permeated with dew, and by taking them back to its burrow where the humidity is relatively high it can improve

the water content further. Water is extracted efficiently from the food by the gerbil's digestive system, thus minimizing water loss in the feces, and retained by the efficient kidneys which produce only a few drops of concentrated urine.

Other features of the gerbil seem to be adaptations to reduce the risk of being captured and killed by predators. All gerbils take the color of the ground on which they live—even subpopulations of a single species living in different habitats: gerbils found on dark lava sands are dark brown whereas members of the same species living on red sand are red. The purpose of the match seems to be to conceal the gerbil from flying predators, though its effectiveness might seem to be compromised by the tail ending in a tuft of contrasting color. This probably acts as a decoy: it distracts a predator and if caught will come away from the gerbil's body with part of or the entire tail.

Another special structure of the gerbil is a particularly large middle ear, which is largest in those species that live in open habitats. It enables the animal to hear sounds of low frequency, such as the wing beats of an owl. The eyes of the gerbil are also positioned so that they give it a wide field of vision. This may also help the animal to become aware of predators.

The geographical range of gerbils can be divided in three major areas. The first is Africa south of the Sahara, mainly savanna and steppes, but also the Namib and Kalahari deserts. Here the temperature does not fall below freezing in the winter. The second area includes the "hot" deserts and semi-desert regions along the Tropic of Cancer in North Africa and Southwest Asia plus Som-

alia, east Ethiopia and north Kenya. The third covers the deserts, semideserts and steppes in Central Asia where winter temperatures fall below freezing.

To a limited extent the genera of gerbils can be formed into groups that inhabit each of these three areas. Gerbils grouped with the *Tatera* species occur in the first area, except for the Antelope rat. The *Gerbillus*-type gerbils occur in the second area, but only the *Meriones* group live in the third area, though some also live in the second area.

Gerbils are basically vegetarians, eating various parts of plants—seeds, fruits, leaves, stems, roots, bulbs etc—but many species will eat anything they encounter, including insects, snails, reptiles and even other small rodents. Some species are very specialized, living on one type of food. The nocturnal *Gerbillus* species often search for wind-blown seeds in deserts. The large Fat sand rat, the Great gerbil and the Antelope rat are basically herbivorous. The Fat sand rat is so specialized that it only occurs where it can find salty succulent plants. The Antelope rat depends on fresh food all year round, so it tends to occur near irrigated crop fields. Wagner's gerbil has such a liking for snails

▼ **Representative species of gerbils.** (**1**) A South African pygmy gerbil (*Gerbillurus paeba*) grooming its muzzle and spreading secretions. (**2**) A Tamarisk gerbil (*Meriones tamariscinus*) exposing its ventral gland. (**3**) A Libyan jird (*Meriones libycus*) making an attack. (**4**) A Cape short-eared gerbil (*Desmodillus auricularis*) making a submissive crouch. (**5**) A Great gerbil (*Rhombomys opimus*) with a heap of sand and feces or urine. (**6**) *Gerbillus gerbillus* (one of the smaller gerbils) marking sand with secretions from its ventral gland. (**7**) A female Mongolian gerbil (*Meriones unguiculatus*) with hair raised darting away from a male (part of the mating sequence). (**8**) A Fat sand rat (*Psammomys obesus*) holding and sniffing a ball of sand and urine.

The Communal Life of the Mongolian Gerbil

Mongolian gerbils live in large social groups which at their largest, in summer, consist of 1–3 adult males, 2–7 adult females and several subadults and juveniles; their home is a single burrow. Detailed studies have demonstrated that they engage in various activities as a group. For example they all collect food to hoard for the winter. They also spend the winter together in their burrow. Their integrity as a community seems, under normal circumstances, to be jealously guarded. Strange gerbils—and other animals—are chased off.

The interesting question that arises from this situation is: who are the parents of the subadults and juveniles? It is not evident from the behavior of males and females within the community, even though they have been observed to form pairs within the groups studied.

They may be the offspring of young adults that migrated from another burrow, but in theory this is unlikely. If these animals were to leave late in the summer to establish their own burrow community elsewhere they would be vulnerable to predators, they would suffer the effects of bad weather, and possibly they would have to contend with other gerbils into whose territories they might wander (when population densities are high there may be as many as 20 burrows per acre, 50 per ha). Moreover they would lose the food they had helped to collect for winter. The most serious objection, however, is that this pattern would perpetuate inbreeding and produce genetic problems.

The solution to the problem has come from the study of animals in captivity. This showed firstly that communal groups do remain stable and territorial, but when females are in heat they leave their own territories and visit neighboring communities to mate. They then return to their own burrows where their offspring will eventually grow up under the protection and care of not their mother and father but of their mother and their uncles.

trast species from savannas, where food is more abundant, are more social: there have been reports of stable pairs being formed and even of family structures emerging.

The most complex social arrangements have been found in those species within the *Meriones* group that live in areas with cold winters. Groups larger than families gather in single burrows, perhaps to keep each other warm, but also perhaps to guard food supplies. The best-known example is the Great gerbil of the Central Asian steppes which lives in large colonies composed of numerous subgroups which themselves have developed from male-female pairs. A similar social structure is found among Mongolian gerbils, but the *Meriones* species that occur in the hot and cooler areas of North Africa and Asia are reported to be solitary in the hot climates but social in the cooler.

Likewise reproduction seems to be linked to climate and food. Desert-dwelling species, like most desert rodents, give birth after the rainy season. Gerbils in areas where fresh food is available may reproduce all the year round, a female perhaps giving birth to two or three litters a year.

The number of young born can vary between one and 12, but the mean lies between three and five according to species. The young appear in a pre-mature state— hairless, eyes closed, unable to regulate their body temperatures. For about 20 days they then depend on their mother's milk. Where there is a breeding season only those born early within it become sexually mature so as to breed in the same season (when aged about two months). Those born later become sexually mature after about six months and breed during the next season.

Most gerbils live in areas of the world uninhabited by man. When the two do come into contact, in the Asian steppes and in India for example, the natural activities of the gerbils make them man's natural enemy. When collecting food to hoard for winter they will pilfer from crops. When burrowing they can cause great damage to pasture, and to irrigation channels, the embankments of road and railways, even to the foundations of buildings. They also carry the fleas that transmit deadly diseases, such as plague and the skin disease Leishmaniasis. Though they serve mankind in medical research, man is keen to eradicate the gerbil when it interferes with his own life. Many gerbil burrows are destroyed, by gasing and plowing, even though some of them may have been used by generations of gerbils for hundreds of years. GA

in its diet that it virtually threatens the existence of local snail populations: big piles of empty shells are found outside this gerbil's burrows. Most gerbils in fact take the precaution of carrying their food back to their burrows and consuming it there. Species that live in areas with cold winters must hoard in order to survive. One Mongolian gerbil was found to have hidden away 44lb (20kg) of seeds in its burrow. The Great gerbil not only hoards plants in its burrow but constructs large stacks outside. Some have been found measuring 3 feet in height and ten in length (1 × 3m).

As yet the social organization of gerbils has been little studied. Species that live in authentic deserts, whatever genera and groups they belong to, tend to lead solitary lives, though burrows are often found to be close enough together that colonies could exist. Perhaps because the supply of food cannot be guaranteed in such an environment each animal fends for itself. By con-

DORMICE

Families: Gliridae, Seleviniidae
Eleven species in 8 genera.
Distribution: Europe, Africa, Turkey, Asia, Japan.

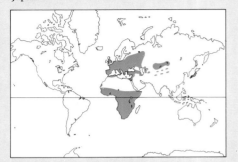

Habitat: wooded and rocky areas, steppe, gardens.

Size: head-body length 2.4–7.5in (6.1–19cm); tail length 1.6–6.5in (4–16.5cm; weight 0.5–7oz (15–200g).

Gestation: 21–32 days.
Longevity: 3–6 years in wild.

Family Gliridae
Species include: **African dormouse** or **Black and white dormouse** (*Graphiurus murinus*); **Common dormouse** or **Hazel mouse** (*Muscardinus avellanarius*); **Edible dormouse** or **Fat dormouse** or **Squirrel-tailed dormouse** (*Glis glis*); **Forest dormouse** or **Tree dormouse** (*Dryomys nitedula*); **Garden dormouse** or **Orchard dormouse** (*Eliomys quercinus*); **Japanese dormouse** (*Glirulus japonicus*); **Mouse-like dormouse** or **Asiatic dormouse** (*Myomimus personatus*).

Family Seleviniidae
Desert dormouse (*Selevinia betpakdalensis*).

THE dormice or Gliridae originated at least as early as the Eocene era (60–40 million years ago). In the Pleistocene era (2 million–10,000 years ago) giant forms lived on some Mediterranean islands. Today dormice are the intermediates, in form and behavior, between mice and squirrels. Key features of dormice are their accumulations of fat and their long hibernation period (about seven months in most European species). The Romans fattened dormice in a special enclosure (Latin *glirarium*) while the French have a phrase "To sleep like a dormouse," similar to the English phrase "To sleep like a log."

Dormice are very agile. Most species are adapted to climbing but some also live on the ground (eg Garden and Forest dormice). The Mouse-like dormouse is the only dormouse that lives only on the ground. The four digits of the forefeet and the five digits of the hind feet have short, curved claws. The underside of each foot is bare with a cushion-like covering. The tail is usually bushy and often long, and in some species (Fat dormouse, Common dormouse, Garden dormouse, Forest dormouse, African dormouse) it can come away when seized by other dormice or predators. Hearing is particularly well developed, as is the ability to vocalize. Fat, Common, Garden and African dormice make use of clicks, whistles and growlings in a wide range of behavior—antagonistic, sexual, explorative, playful.

Dormice are the only rodents that do not have a cecum, which indicates a diet with little cellulose. Analysis of the contents of the stomachs of dormice has shown that they are omnivores whose diet varies according to season and between species according to region. The Edible and Common dormice are the most vegetarian, eating quantities of fruits, nuts, seeds and buds. Garden, Forest and African dormice are the most carnivorous—their diets include insects, spiders, earthworms, small vertebrates, but also eggs and fruit. In France 40–80 percent of the diet of the Garden dormouse consists of insects, according to region and season. But there is also another factor at work. In summer the Garden dormouse eats mainly insects and fruit, in fall little except fruit, even though the supply of insects at this time of year is plentiful. The change in the content of the diet is part of the preparation for entering hibernation; the intake of protein is reduced, sleep is induced.

In Europe dormice hibernate from October to April with the precise length varying between species and according to region.

During the second half of the hibernation period they sometimes wake intermittently, signs of the onset of the hormone activity that stimulates sexual activity. Dormice begin to mate as soon as they emerge from hibernation, females giving birth from May onwards through to October according to age. (Not all dormice that have recently become sexually mature participate in mating.) The Edible and Garden dormice produce one litter each per year but Common and Forest dormice can produce up to three. Vocalizations play an important part in mating. In the Edible dormouse the male emits calls as he follows the female, in the Garden dormouse the female uses whistles to attract the male. Just before she is due to give birth the female goes into hiding and builds a nest, usually globular in shape and located off the ground, in a hole in a tree or in the crook of a branch for example. Materials used include leaves, grass and moss. The Garden and Edible dormice use hairs and feathers as lining materials. The female Garden dormouse scent marks the

▷ **Clinging tight.** OVERLEAF Most Garden dormice in fact live in forests across central Europe, though some inhabit shrubs and crevices in rocks.

◁ **Coming down,** a Common dormouse. This richly colored species lives in thickets and areas of secondary growth in forests. It is particularly fond of nut trees.

◁ **Looking out,** BELOW LEFT, an Edible dormouse. This squirrel-like dormouse also lives in trees, but also inhabits burrows. It has a great liking for fruit.

▽ **Curled up,** a Common dormouse in hibernation. For the long winter sleep this dormouse resorts to a nest, either in a tree stump, or amidst debris on the ground, or in a burrow. The length of hibernation is related to the climate, but can last for as long as nine months.

area around the nest and defends it.

Female dormice give birth to between two and nine young, with four being the average litter size in about all species. The young are born naked and blind. In the first week after birth they become able to discriminate between smells, though the exchange of saliva between mother and young appears to be the means whereby mother and offspring learn to recognize each other. This may also aid the transition from a milk diet to a solid food one. At about 18 days young become able to hear and at about the same time their eyes open. They become independent after about one month to six weeks. Young dormice then grow rapidly until the time for hibernation approaches, when their development slows. Sexual maturity is reached about one year after birth, towards the end of or after the first hibernation.

Dormice populations are usually less dense than those of many other rodents. There are normally between 0.04 and 4 dormice per acre (0.1–10 per ha). They live in small groups, half of which are normally juveniles, and each group occupies a home range, the main axis of which can vary from 330ft (100m) in the Garden dormouse to 660ft (200m) in the Edible dormouse. In urban areas the radiotracking of Garden dormice has indicated that their home range is of an elliptic shape and related to the availability of food. In fall the home range is about 10,800sq ft (about 1,000sq m). A recent study of the social organization of the Garden dormouse has revealed significant changes in behavior in the active period between hibernations. In the spring, when Garden dormice are emerging from hibernation, males form themselves into groups in which there is a clearcut division between dominant and subordinate animals. As the groups are formed some males are forced to disperse. Once this has happened, although groups remain cohesive, behavior within them becomes more relaxed so that by the end of the summer the groups have a family character. In the fall social structure includes all categories of age and sex. Despite the high rate of renewal of its members a colony can continue to exist for many years.

The Desert dormouse, which is placed in a family of its own, occurs in deserts to the west and north of Lake Balkhash in eastern Kazakhstan, Central Asia. It has exceptionally dense, soft fur, a naked tail, small ears, and sheds the upper layers of skin when it molts. It eats invertebrates, such as insects and spiders and is mostly active in twilight and night. It probably hibernates in cold weather.   CB

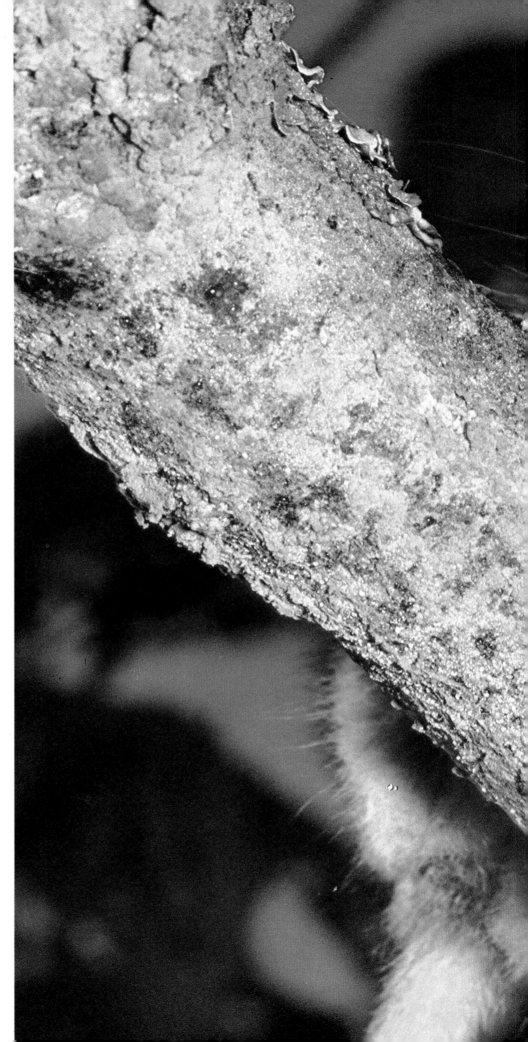

JUMPING MICE, BIRCHMICE AND JERBOAS

Family: Zapodidae
Jumping mice and birchmice
Fourteen species in 4 genera belonging to 2 subfamilies.
Distribution: N America and Eurasia.

Family: Dipodidae.
Jerboas
Thirty species in 11 genera belonging to 3 subfamilies.
Distribution: N Africa, Turkey, Middle East, C Asia.

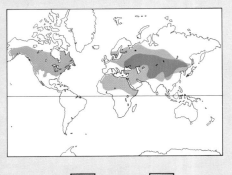

Jumping mice
and birchmice   Jerboas

▼ **Mouse on stilts,** a Meadow jumping mouse. Its most important food is grass seeds, many of which are picked from the ground, but some are pulled from grass stalks. To eat timothy seeds, for example, the mouse may climb up the stalk and cut off the seed heads. At other times the mouse will reach as high as it can, cut off the stalk, pull the top portion down, cut it off again, etc, until the seed head is reached. The stem, leaf stalk and uneaten seeds are usually left in a criss-cross pile of match-length parts. Another favorite seed is that of the touch-me-not. These seeds taste like walnut, but have bright turquoise endosperm which turns the entire contents of the stomach brilliant blue.

THE name "jumping mouse" is something of a misnomer. All **jumping mice** are equipped for jumping, with long back feet, and long tails to help them maintain their balance in the air, but the most common species, those belonging to the genus *Zapus*, are more likely to crawl under vegetation or to run by making a series of short hops rather than long leaps. However, the Woodland jumping mouse often moves by bounding 5–10ft (1.5–3m) at a time. In addition to the feet and tail, the outstanding characters of jumping mice are their colorful fur and their grooved upper incisors. The function of the groove is unknown: it may improve the efficiency of the teeth as cutting tools, or it may just strengthen them.

Jumping mice are not burrowers. They live on the surface of the ground, though their nests may be underground or in a hollow log or other protected places, and the hibernating nest is often at the end of a burrow in a bank or other raised area. For the most part, however, jumping mice hide by day in clumps of vegetation. They also usually travel about in thick herbaceous cover, though they will use runways or sometimes burrows of other species when present.

The habitat of jumping mice varies, but lush grassy or weedy meadows are the preferred habitat of Meadow jumping mice, although they are often quite abundant in wooded areas, in patches of heavy vegetation (especially of touch-me-not, *Impatiens*), particularly in areas where there are no Woodland jumping mice. Woodland jumping mice usually occur in woods, almost never in open areas, and are most abundant in wooded areas with heavy ground cover. Moisture is often mentioned as a factor favorable to jumping mice, but it seems more important as a factor favoring the development of lush vegetation rather than as a factor directly favoring the mice.

Jumping mice are profound hibernators, hibernating for 6–9 months of the year according to species, locality and elevation. The Meadow jumping mouse in the eastern USA usually hibernates from about October to late April. Individuals that hibernate successfully put on about 0.21–0.35oz (6–10g) of fat in the two weeks prior to entering hibernation. This they do by sleeping for increasingly longer periods until they attain deep hibernation with their body temperature just a little above freezing. Their heart rate, breathing rate and all bodily functions drop to low levels. However, the animals wake about every two weeks, perhaps urinate, then go back to sleep. In the spring the males appear above ground about two weeks before the females. Of the animals active in the fall, only about a third—the larger ones—are apparently able to put on the layer of fat, enter hibernation and awaken in the spring. The remainder—young individuals or those unable to put on adequate fat—apparently perish during the winter retreat.

Jumping mice give birth to their young in a nest of grass or leaves either underground, in a clump of vegetation, in a log or in some other protected place. Gestation takes about 17 or 18 days, or up to 24 if the female is lactating. Each litter contains about 4–7 young. Litters may be produced at any time between May and September, but most enter the world in June and August. Most females probably produce one litter per year.

Meadow jumping mice eat many things, but seeds, especially those from grasses, are the most important food. The seeds eaten change with availability.

The major animal foods eaten by jumping mice are moth larvae (primarily cutworms) and ground and snout beetles. Also important in the diet is the subterranean fungus *Endogone*. This forms about 12 percent of the diet (by volume) in the meadow jumping mice, and about 35 percent of the diet in the Woodland jumping mouse.

Birchmice differ from jumping mice in having scarcely enlarged hind feet and upper incisors without grooves. Moreover their legs and tail are shorter than those of jumping mice, yet they travel by jumping and climb into bushes using their outer toes to hold on to vegetation and their tails for partial support. Birchmice, also unlike jumping mice, dig shallow burrows and make nests of herbaceous vegetation underground.

Like jumping mice, birchmice are active primarily by night. They can eat extremely large amounts of food at one time and can also spend long periods without eating. Their main foods are seeds, berries and insects. Birchmice hibernate in their underground nests for about half of the year. It has been suggested that *Sicista betulina* spends the summer in meadows but hibernates in forest. Gestation probably lasts about 18–24 days and parental care for another four weeks. Studies of *S. betulina* in Poland have shown that one litter a year is produced and that any female produces only two litters during her lifetime.   JOW

Jerboas are small animals built for jumping. Their hind limbs are elongated—at least four times as long as their front legs—and in most species the three main foot bones are fused into a single "cannon bone" for greater strength (in the subfamilies Dipodinae and Euchoreutinae, but not in the Cardiocraniinae). The outer toes on the hind feet are small in size and do not touch the ground in species with five toes. In other species the outer toes are absent, so there are three toes on each hind foot. One species, *Allactaga tetradactylus*, has four toes. Jerboas living in sandy areas have tufts of hairs on the undersides of the feet which serve as snowshoes on soft sand and help them to maintain traction and to kick sand backwards when burrowing. These jerboas also have tufts of hair to help keep sand out of the ears.

Some jerboas, those belonging to the genus *Jaculus* for example, have a fold of skin which can be pulled forward over the nostrils when burrowing. Jerboas use their long tails as props when standing upright and as balancing organs when jumping. Jerboas are nocturnal and have large eyes.

The well-developed jumping ability of jerboas enables them to escape from predators as well as to move about, though they also move by slow hops. Otherwise only the hind legs are used in moving; the animal walks on its hind legs. The front feet then can be used for gathering food. Jumps of 5–10ft (1.5–3m) are used when the animal moves rapidly. Desert jerboas, *Jaculus*, can jump vertically to nearly 3ft (1m).

Jerboas feed primarily on seeds, but sometimes also on succulent vegetation. In some areas they may be a pest to growers of melons. They also eat insects, and one species, *Allactaga sibirica*, feeds primarily on beetles and beetle larvae. One individual of *Salpingotus* in captivity ate only invertebrates. In *Dipus* all individuals in a population emerge for their nightly forays at about the same time, and move by long leaps to their feeding grounds, which may be some distance away. There they feed on plants, especially those with milky juices, but they also smell out underground sprouts and insect larvae in underground galls (gallnuts). Like pocket mice, jerboas do not drink water; they manufacture "metabolic water" from food.

Some Jerboas hibernate during the winter, surviving off their body fat. Also, some species enter torpor during hot or dry periods. They are generally quiet, but when handled will sometimes shriek or make grunting noises. Some species have been known to tap with a hind foot when inside their burrows.

In northern species mating first occurs shortly after the emergence from hibernation, but most female jerboas probably breed at least twice in a season, producing litters of between two and six young.

There are four kinds of burrows used by various jerboas, depending on their habits and habitats: temporary summer day burrows for hiding during the day, temporary summer night burrows for hiding during nightly forays, permanent summer burrows used as living quarters and for producing young, and permanent winter burrows for hibernation. The two temporary burrows are simple tubes, which are in length respectively 8–20in (20–50cm) and 4–8in (10–20cm).

The permanent summer burrows have secondary chambers for food storage, and the permanent winter burrows are at least 9in (22cm) below the surface and also have secondary chambers. Some species build a mound at the entrance, and some provide one or more accessory exits. The Comb-toed jerboa lives in sand dunes where it burrows into the protected side of the dunes. Most of the burrows that have been dug up consisted of single passages. This is one species from which tapping sounds have been recorded from within the burrow.　JOW

▲ **The long-legged burrow-dweller.** The burrows used by desert jerboas can run as deep as 6ft (about 2m). At this depth the animals are insulated against fluctuations in outside temperature. Burrows can be complex, with passages off the main chambers from which the animals can "burst" through the soil to the surface when disturbed or threatened. There is usually one jerboa to a burrow, but the desert jerboas (as here) are fairly sociable and live in loose colonies with two or three animals often using the same nest. In spite of having long hind legs desert jerboas are adept at burrowing, using their short front feet and incisor teeth for digging, and their hind feet for throwing away the sand.

Jumping mice and birchmice

Subfamily Zapodinae
Jumping mice

Four species in three genera. Distribution: N America with one species in China (*Eozapus setchuanus*). Habitat: meadows, moors, steppes, thickets, woods. Size: head-body length 3–4.3in (7.6–11cm), tail length 5.9–6.5in (15–16.5cm), weight up to 1oz (28g). Gestation: about 17–21 days in *Zapus* and *Napaeozapus*. Longevity: probably one or two years. Species: *Eozapus setchuanus*. *Napaeozapus insignis* (**Woodland jumping mouse**). *Zapus hudsonius* (**Meadow jumping mouse**), *Z. princeps*.
(The animal sometimes classified separately as *Z. trinotatus* is here considered to be synonymous with *Z. princeps*.)

Subfamily Sicistinae
Birchmice

Nine species in one genus. Distribution: Eurasia. Habitat: mainly birch forests but other habitats also. Size: head-body length 1.9–3.5in (5–9cm), tail length 2.6–3.9in (6.5–10cm), weight up to 1oz (28g). Gestation: 18–24 days in *Sicista betulina*. Longevity: probably less than a year. Species of the genus *Sicista* include *S. betulina* (N Eurasia), *S. caucasica* (W Caucasus and Armenia), *S. concolor* (China), *S. subtilis* (USSR and E Europe).

Jerboas

Habitat: desert, semidesert, steppe, including patches of bare ground. Size: head-body length 1.6–10in (4–26cm), tail length 2.7–11in (7–30cm), hind foot length 0.8–10in (2–4cm). Gestation: 25–42 days. Longevity: probably less than two years.

Subfamily Cardiocraniinae

Genera: **Five-toed dwarf jerboa** (*Cardiocranius paradoxus*), W China, Mongolia. *Salpingotulus michaelis* Pakistan. **Three-toed dwarf jerboas** (Genus *Salpingotus*) Asia in deserts. Three species.

Subfamily Dipodinae

Genera: **four-** and **five-toed jerboas** (Genus *Allactaga*) NE Africa (Libyan desert), Arabian peninsula, C Asia. Eleven species. **Lesser five-toed jerboa** (*Alactagulus pumilio*) S European Russia in clay and salt deserts and prairies. **Feather-footed jerboa** (*Dipus sagitta*) China and USSR. **Desert jerboas** (Genus *Jaculus*) N Africa, Russia, Iran, Afghanistan, Pakistan in various habitats (including desert, rocky areas, meadows). Five species. **Comb-toed jerboa** (*Paradipus ctenodactylus*) USSR in dry, sandy deserts. **Fat-tailed jerboas** (Genus *Pygeretmus*) USSR in salt and clay deserts. Three species. **Thick-tailed three-toed jerboa** (*Stylodipus telum*) Russia, Mongolia, China, in clay and gravel deserts.

Subfamily Euchoreutinae

Genera: **Long-eared jerboa** (*Euchoreutes naso*) China and Mongolia in sandy areas.

CAVY-LIKE RODENTS

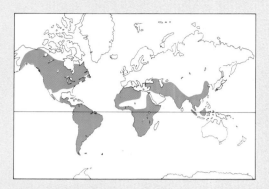

Suborder: Caviomorpha
Eighteen families: 60 genera: 188 species.
Distribution: America, Africa, Asia.

Habitat: desert, grassland, savanna, forest.

Size: head-body length from 6.8in (17cm) in gundis to 53in (134cm) in the capybara; weight from 6.2oz (175g) in Speke's gundi to 141lb (64kg) in the capybara.

New World porcupines
Family: Erethizontidae
Ten species in 4 genera.

Cavies
Family: Caviidae
Fourteen species in 5 genera.

Capybara
Family: Hydrochoeridae
One species.

Coypu
Family: Myocastoridae
One species.

Hutias
Family: Capromyidae
Thirteen species in 4 genera.

Pacarana
Family: Dinomyidae
One species.

Pacas
Family: Agoutidae
Two species in 1 genus.

Agoutis and acouchis
Family: Dasyproctidae
Thirteen species in 2 genera.

Chinchilla rats
Family: Abrocomidae
Two species in 1 genus.

Spiny rats
Family: Echimyidae
Fifty-five species in 15 genera.

Chinchillas and viscachas
Family: Chinchillidae
Six species in 3 genera.

Degus or Octodonts
Family: Octodontidae
Eight species in 5 genera.

Tuco-tucos
Family: Ctenomyidae
Thirty-three species in 1 genus.

Cane rats
Family: Thryonomyidae
Two species in 1 genus.

THE familiar Guinea pig is a representative of a large group of South American rodents usually referred to as the caviomorph rodents and formally classified in a suborder, Caviomorpha. Most are large rodents, confined to South and Central America. Although they are very diverse in external appearance and are generally classified in separate families, they share sufficient characteristics to make it virtually certain that they constitute a natural, inter-related group.

External many caviomorphs have large heads, plump bodies, slender legs and short tails, as in the guinea pigs, the agoutis and the giant capybara, the largest of all rodents at over 3 feet (about 1m) in length. Others, however, eg some spiny rats of the family Echimyidae, come very close in general appearance to the common rats and mice.

Internally the most distinctive character uniting these rodents is the form of the masseter jaw muscles, one branch of which extends forwards through a massive opening in the anterior root of the bony zygomatic arch to attach on the side of the rostrum. At its other end it is attached to a characteristic outward-projecting flange of the lower jaw. Caviomorphs are also characterized by producing small litters after a long gestation period, resulting in well-developed young. Guinea pigs, for example, usually have two or three young after a gestation of 50–75 days, compared with seven or eight young after only 21–24 days in the Norway (murid) rat.

This group is also frequently referred to as the hystricomorph rodents, implying that the various Old World porcupines, family Hystricidae, are closely related to the American "caviomorphs." Although they share all the caviomorph features mentioned above there has been considerable debate as to whether such features indicate a common ancestry or that they have evolved independently in two groups—a problem related to the question as to whether the caviomorphs reached South America from North America or from Africa. The African cane rats (family Thryonomyidae) are closely related to the Hystricidae but some other families, namely the gundis (Ctenodactylidae) and the rock rat (Petromyidae) are much more doubtfully related, showing only some of the "caviomorph" characters.

Most of the American caviomorphs are terrestrial and herbivorous but a minority, the porcupines (Erethizontidae), are arboreal and one group, the tuco-tucos (Ctenomyidae), are burrowers.   GBC

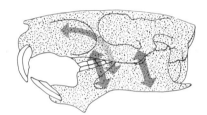

▲ **Distinguishing feature of cavy-like rodents.** The deep masseter muscle (blue) provides the gnawing action, extending forward through an opening in the zygomatic arch to attach to the muzzle. The lateral masseter (green) is only used in closing the jaw.

▶ **The face of a true cavy-like rodent.** Like most South American members of this suborder the degu has a plump, well-furred body and a large head.

▼ **A distant relative:** an Old World porcupine (Cape porcupine). Even porcupine young are born with quills.

African rock rat
Family: Petromyidae
One species.

Old World porcupines
Family: Hystricidae
Eleven species in 4 genera.

Gundis
Family: Ctenodactylidae
Five species in 4 genera.

African mole-rats
Family: Bathyergidae
Nine species in 5 genera.

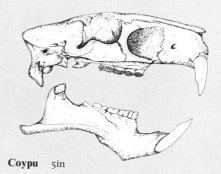

 Coypu　5in

Coypu

Mara

Capybara

Tuco-tuco

Skulls and teeth of cavy-like rodents.

Most cavy-like rodents have rather angular skulls and very strongly developed incisor teeth. The wearing surfaces of the four cheekteeth show enormous variation in pattern and complexity amongst the different species. Those of the coypu are typical of a large group of herbivorous species, including the agoutis and the American porcupines, and are closely paralleled in the Old World porcupines and cane rats. The teeth of the mara and of the capybara, although superficially very different in degree of complexity, resemble each other in being ever-growing as in the unrelated but also grass-eating voles and rabbits. At the other extreme the tuco-tucos have surprisingly simple cheekteeth considering that they feed mainly on roots and tubers.

NEW WORLD PORCUPINES

Family: Erethizontidae
Eleven species in 5 genera.
Distribution: N America (except SE USA),
S Mexico, C America, N S America.

Habitat: forest areas, open grasslands, desert,
canyon.

Size: ranges from head-body length 12in
(30cm) and weight 32oz (900g) in the
prehensile-tailed porcupines to head-body
length 34in (86cm) and weight 40lb (18kg) in
the North American porcupine.

Gestation: 210 days in the North American
porcupine.

Longevity: up to 17 years in the North
American porcupine.

Prehensile-tailed porcupines
Genus *Coendou*
S Panama, Andes from NW Colombia to N
Argentina, NW Brazil; forest areas. Two
species: **Prehensile-tailed porcupine**
(*C. prehensilis*), **South American tree porcupine**
(*C. bicolor*).

Upper Amazonian porcupine
Echinoprocta rufescens
C Colombia; forest areas.

North American porcupine
Erethizon dorsatum
Alaska, Canada, USA (except extreme SW, SE
and Gulf coast states), N Mexico; forest areas.

South American tree porcupines
Genus *Sphiggurus*
S Mexico, C America, S America as far S as
N Argentina; forest areas. Six species including:
Mexican tree porcupine (*S. mexicanus*), **South
American tree porcupine** (*S. spinosus*).

Thinned-spined porcupine
Chaetomys subspinosus
E and N Brazil, forest areas.

NEW World porcupines strongly resemble Old World porcupines, but New World porcupines are arboreal in habits, unlike Old World porcupines which are terrestrial. For such heavy-bodied animals they are excellent climbers with well-developed claws and unfurred soles on their large feet. The soles consist of pads and creases which increase the supporting surface and the gripping power of the feet. Individual genera have further modifications to improve their climbing abilities. The prehensile-tailed porcupines and the South American tree porcupines—the most arboreal genera—have smaller first digits on their hind feet than the other genera, but they are incorporated in the footpads which increases the width and the gripping power of the pads.

The same genera also have long spineless tails for grasping. Their tips form upward curls and have a hard skin (callus) on the upper surface. In the prehensile-tailed porcupines the tail contributes 9 percent of the total body weight; nearly half of the weight of the tail is composed of muscle fibers.

New World porcupines are very nearsighted, and have keen senses of touch, hearing and smell. They produce a variety of sounds—moans, whines, grunts, coughs, sniffs, shrieks, barks and wails. All porcupines have large brains and appear to have good memories.

In habits porcupines range from the North American porcupine, which is semi-arboreal, to prehensile-tailed and South American tree porcupines, which are specialized, arboreal feeders. All forms spend much of their time in trees, but even tree porcupines are known to come to the ground to feed and to move from one tree to another. In winter North American porcupines feed on conifer needles and on the bark of a variety of trees, except Red maple, White cedar and hemlock. During the summer these porcupines feed more frequently on the ground and select roots, stems, leaves, berries, seeds, nuts and flowers. In the spring they often come out from forested areas into meadows to feed on grasses in the evening hours. They will eat bark at all times of the year, and can be destructive to forest plantations. Prehensile-tailed and South American tree porcupines feed more on leaves and have many characteristics of an arboreal leaf-eater. However, they are also reported to feed on tender stems, fruits, seeds, roots, tubers, insects and even small reptiles.

In the North American porcupines the female reaches sexual maturity when about 18 months old. The estrous cycle is 29 days,

and these animals may have more than one period of estrus in a year. They have a vaginal closure membrane so females form a copulatory plug. The gestation period averages 210 days, and in both North American and prehensile-tailed porcupines usually one young is produced (rarely twins). The weight of the precocial newborn is about 14oz (400g) in prehensile-tailed porcupines and 21oz (600g) in North American porcupines. Lactation continues for 56 days, but the animals also feed on their own after the first few days. Porcupine young are born with their eyes open and are able to walk. They exhibit typical defensive reactions and within a few days are able to climb trees. These characteristics probably explain why infant mortality is very low. Porcupines grow for three or four years before they reach adult body size.

The home range of North American porcupines in summer averages 36 acres

▲ **Sustained on a branch.** The Prehensile-tailed porcupines live mainly in the middle and upper layers of forests in Central and South America, only descending to the ground to eat. Although their claws are large, firm and stiff, they do not pin themselves to or cling to trees but rely for adhesion on their weight, a firm hold with their claws and, as seen here on the far left, their tail which can be coiled around branches and has a callus pad to provide grip.

▶ **In search of food,** a South American tree porcupine. More time is spent on the ground in summer than in winter. The prehensile-tailed porcupines are more herbivorous than other genera, eating large quantities of leaves, roots, stems, blossoms and fruit. Others have more developed tastes for insects and small reptiles.

(14.6ha). In winter, however, they do not range great distances—they stay close to their preferred trees and shelters. Prehensile-tailed porcupines can have larger ranges, though these vary from 20 to 94 acres (8–38ha). They are reported to move to a new tree each night, usually 660–1,300ft (200–400m) away, but occasionally up to 2,300ft (700m). Prehensile-tailed porcupines in South Guyana are known to reach densities of 130–260 individuals per sq mi (50–100 per sq km). They have daily rest sites, in trees, 20–33ft (6–10m) above the ground, usually on a horizontal branch. These porcupines are nocturnal, change locations each night and occasionally move on the ground during the day. Male prehensile-tailed porcupines are reported to have ranges of up to four times as large as those of females.

Porcupines in general are not endangered, and the North American porcupine can in fact be a pest. The fisher (a species of marten) has been reintroduced to some areas of North America to help control porcupines, one of its preferred prey. The fisher is adept at flipping the North American porcupine over so that its soft and generally unquilled chest and belly are exposed. The fisher attacks this area, killing and eating the porcupine from below. One study found that porcupines declined by 76 percent in an area of northern Michigan (USA) following the introduction of the fisher. Prehensile-tailed porcupines are frequently used for biomedical research, which contributes to the problem of conserving the genus, but the main threat is habitat destruction. In Brazil prehensile-tailed porcupines have been affected by the loss of the Atlantic forest, and the South American tree porcupine is included on the list of endangered species published by the Brazilian Academy of Sciences. One species of porcupine may have become extinct in historic times: *Sphiggurus pallidus*, reported in the mid-19th century in the West Indies, where no porcupines now occur.   CAW

▶ **Lord of the conifer forest.** This is the North American porcupine, found in forests across most of Canada, the USA and northern Mexico, but which is mainly terrestrial. It has relatively poor eyesight, cannot jump, moves slowly and clumsily, but frequently climbs trees to enormous heights, in search of food—twigs, leaves, berries, nuts. Its small intestine digests cellulose efficiently.

▼ **Almost a primate**—a North American porcupine in Alaska.

CAVIES

Family: Caviidae
Fourteen species in 5 genera.
Distribution: S America (mara in C and S Argentina).

Habitat: open areas in forests, semiarid thorn scrub, arboreal savanna, Chaco, pampas, high altitude puna, desert (scrub desert and grasslands for mara).

Size: in small cavies ranges from head-body length 8.7in (22cm), weight 10.7oz (300g) in the genus *Microcavia* to head-body length 15in (38cm), weight 2.2lb (1kg) in the genus *Kerodon*: in mara head-body length 19.7–30in (50–75cm), tail length 1.8in (4.5cm), weight 17.6–19.8lb (8–9kg).

Gestation: in small cavies varies from 50 days in genera *Galea* and *Microcavia* to 75 days in genus *Kerodon*; in mara 90 days.

Longevity: in small cavies 3–4 years (up to 8 in captivity); in mara maximum 15 years.

Guinea pigs and cavies
Genus *Cavia*.
S America, in the full range of habitats.
Coat: grayish or brownish agouti; domesticated forms vary. Five species: *C. aperea, C. fulgida. C. nana, C. porcellus* (**Domestic guinea pig**), *C. tschudii.*

Mara or Patagonian hare
Genus *Dolichotis*.
S America (C and S Argentina), occurring in open scrub desert and grasslands. Head and body are brown, rump is dark (almost black) with prominent white fringe round the base; belly is white. Two species: *D. patagonum* and *D. salinicolum.*

Genus *Galea*
S. America, in the full range of habitats.
Coat: medium to light brown agouti with grayish-white underparts. Three species: *G. flavidens, G. musteloides, G. spixii.*

Rock cavy
Kerodon rupestris.
NE Brazil, occurring in rocky outcrops in thorn-scrub.
Coat: gray, grizzled with white and black; throat is white, the belly yellow-white, the rump and backs of the thighs reddish.

Desert cavies
Genus *Microcavia*.
Argentina and Bolivia, in arid regions.
Coat: a coarse dark agouti, brown to grayish.
Three species: *M. australis, M. niata, M. shiptoni.*

Most people are familiar with cavies, but under a different and somewhat misleading name: the Guinea pig. "Guinea" refers to Guyana, a country where cavies occur in the wild; and the short, squat body gives this rodent a piggish appearance, at least to the imaginative eye. The pork-like quality of the flesh doubtless contributes to the use of the name.

The Guinea pig of pet stores and laboratories has little in common with its wild cousins besides a shared evolutionary history. The Domestic guinea pig was being raised for food by the Incas when the conquistadors arrived in Peru in the 1530s and is now found the world over with one exception: it no longer occurs in the wild. Wild cavies share the same squat body form as the Guinea pig, but their simple external appearance belies their ecological adaptability. Cavies are among the most abundant and widespread of all South American rodents.

All cavies (except the mara or Patagonian cavy, for which see pp110–111) share a basic form and structure. The body is short and robust, the head is large, contributing about one-third of the total head-body length. The eyes are fairly large and alert, the ears large but close to the head. The fur is coarse and easily shed when the animal is handled. There is no tail. The forefeet are strong and flat, usually with four digits with sharp claws. The hind feet, with three clawed digits, are elongated markedly. They walk on their soles with the heels touching the ground. The incisors are short, and the cheekteeth, which are arranged in rows that converge towards the front of the mouth, have the shape of prisms and are evergrowing. Both sexes are alike, apart from each possessing certain specialized glands.

Cavies are very vocal, making a variety of chirps, squeaks, burbles and squeals. One genus, *Kerodon*, emits a piercing alarm whistle when frightened. *Galea* rapidly drum their hind feet on the ground when anxious.

Cavies first appeared in the mid-Miocene era of South America. Since their appearance, some 20 million years ago, the family Caviidae has undergone an extensive adaptive radiation, reaching peak diversity 5–2 million years ago, during the Pliocene (when there were 11 genera), then declining in the number of species to present levels (5 genera) during the Pleistocene (about 1 million years ago).

The 12 remaining species of the subfamily Caviinae (all cavies except *Dolichotis* species) are widely distributed throughout South America. All 12 species are to a degree specialized for exploiting open habitats. Cavies can be found in grasslands and scrub forests from Venezuela to the Straits of Magellan. But each cavy genus has also evolved to be able to exist in a slightly different habitat. *Cavia* is the genus most restricted to grasslands. In Argentina *Cavia aperea* is restricted to the humid pampas in the northeastern provinces. *Microcavia australis* is the desert specialist and is found throughout the arid Monte and Patagonian deserts of Argentina. Other *Microcavia* species, *M. niata* and *M. shiptoni*, occur in the arid high-altitude puna (subalpine zone) of Bolivia and Argentina. The specialized genus *Kerodon* is found only in rock outcrops called *lajeiros* which dot the countryside in the arid thorn scrub, or *caatinga*, of northeastern Brazil. *Galea* seem to be the jacks-of-all-trades of the cavies. *Galea* species are

▲ **Hutch in the wild.** Cavy genera occupy a variety of habitats. These include rocky outcrops in which the Rock cavy is common.

▶ **Grazing stock for the table.** The Domestic guinea pig has been bred by South American Indians for its meat for at least 3,000 years.

◀ **Rock cavy mating procedure.** An adult male blocks the path of a female (1), passes under her chin (2) and begins to mount (3).

▶ **Elegance attained.** OVERLEAF Selective breeding has produced Domestic guinea pigs of numerous strains. As pets they are bred for beauty, though in South America some have joined their ancestors in the wild.

found in all of the above habitats, and it is the only genus that occurs along with the other three genera.

Regardless of habitat, all cavies are herbivorous. *Galea* and *Cavia* feed on herbs and grasses. *Microcavia* and *Kerodon* seem to prefer leaves—both genera are active climbers. *Kerodon* are especially surprising because they lack both claws and a tail, two adaptations usually associated with life in trees. The sight of a 28oz (800g) guinea pig scooting along a pencil-thin branch high in a tree is quite striking.

All cavies become sexually mature early, at between one and three months of age. Of the species studied to date, only *Microcavia* show marked seasonality in reproduction. The gestation period in cavies is fairly long for rodents, varying from 50 days (*Galea* and *Microcavia*) to 60 (*Cavia aperea*) and 75 (*Kerodon*). Litter sizes are small, averaging about three for *Galea* and *Microcavia*, two for *Cavia* and 1.5 for *Kerodon*. Young are born highly precocial. Males contribute little obvious parental care. The male generally ignores the female and her young once the litter is born. Whether or not the male defends resources needed by the female is unclear. This seems to be the case with *Kerodon* but not with *Galea*.

With cavies being so similar in form and in diet and reproductive biology, what are the differences between species? Some of the most interesting originate in social behavior and in adaptations, in very subtle ways, to the environment.

Three species of cavies have been studied in northeastern Argentina: *Microcavia australis*, *Galea musteloides* and *Cavia aperea*. *Cavia* and *Microcavia* never occur in the same area, *Cavia* preferring moist grasslands and *Microcavia* more arid habitats. *Galea* occurs with both genera. Competition between *Galea* and *Microcavia* seems to be minimized by different foraging tactics: *Microcavia* is more of a browser, and arboreal. The degree to which *Cavia* and *Galea* interact within the same areas is unknown. Home-range sizes are known only for *Microcavia*, on average 34,500sq ft (approx. 3,200sq m) and for *Galea* 14,000sq ft (approx. 1,300sq m).

All three of these cavies have a similar social structure: mating is promiscuous, no male-female bonds are formed, and there is no permanent social group. But there are some subtle differences. *Microcavia australis*, the species most adapted to arid regions with limited resources, has the highest level of amicable behavior among individuals. *Cavia aperea* occurs in habitats that have a high productivity of grasses and herbaceous vegetation. Although food resources are abundant, *Cavia* is the most aggressive of the cavy genera (adults are especially aggressive towards juveniles, causing them to disperse early). *Galea* cavies live in areas where resources are intermediate in abundance. They have a social structure with moderate levels of adult–juvenile aggression.

Two species of cavies coexist in northeastern Brazil: *Kerodon rupestris* and *Galea spixii*. *Galea spixii* is similar to the Argentine *Galea* in morphology, color, ecology and behavior. They inhabit thorn forests, are grazers and have a noncohesive social organization. *Kerodon rupestris* (the Rock cavy) is markedly different from all the other small cavy species. It is larger and leaner, and has a face that is almost dog-like. All small cavies except *Kerodon* have sharply clawed digits; *Kerodon* have nails growing from under the skin with a single grooming claw on the inside hind toes. Their feet are extensively padded. The modifications of the feet facilitate movement on slick rock surfaces. They are strikingly agile as they leap from boulder to boulder, executing graceful mid-air twists and turns. They are also exceptional climbers, and forage almost exclusively on leaves in trees. There is little competition for resources with *Galea*.

Perhaps the most interesting difference between *Kerodon* and *Galea* is behavioral. *Galea*, like the Argentine cavies, inhabit a relatively homogeneous habitat: the thorn scrub forest. *Galea spixii* also has a promiscuous mating system. *Kerodon* inhabit isolated patches of boulders, many of which can be defended by a single male, and have what appears to be a harem-based mating system: one in which a single male has exclusive access to two or more females. The clumped distribution of the boulder piles allows for the males to be able to do this. This makes *Kerodon* similar to the unrelated hyraxes of eastern Africa. Hyraxes live in kopjes, a kind of rock outcrop similar to the ones occupied by *Kerodon*.

Most cavy species can be found in altered and disturbed habitat, and some do well in and among human habitation. The one exception is *Kerodon*, a unique mammal, which has a patchy distribution throughout its range: Rock cavies are hunted extensively, and are declining in numbers. This specialized rodent is in desperate need of protection. Because these rodents occur in such a rare and patchily distributed habitat, large areas will be needed to assure the existence of *Kerodon*; indeed two research reserves have been set aside in Brazil. TEL

Life-long Partners

Colonial breeding in the monogamous mara

Dawn broke across the Patagonian thornscrub as a large female rodent, with the long ears of a hare and the body and long legs of a small antelope, cautiously approached a den, followed closely by her mate. They were the first pair to arrive, and so walked directly to the mouth of the den. At the burrow's entrance the female made a shrill, whistling call, and almost immediately eight pups of various sizes burst out. The youngsters were hungry, not having nursed since the previous night, and all thronged around the female, trying to suckle. Under this onslaught she jumped and twirled to dislodge the melee of unwelcome mouths which sought her nipples. The female sniffed each carefully, lunging at and chasing off those that were not her own. Finally, she managed to select her own two offspring from the hoard and led them 33ft (10m) away from the den to a site where, despite intermittent harassment from other hungry infants, they would be nursed on and off for an hour or more.

In the meantime her mate sat alert nearby. If another adult pair had approached the den while his female was there, coming to tend their own pups, he would have made a vigorous display directly in front of his mate. If the newcomers had not moved away he would have dashed towards them, with his head held low and neck outstretched, and chased them off. The second pair would then have waited, alert or grazing, at a distance of 65–100ft (about 20–30m). When the original pair had left the area the new pair would then approach the den, to collect their own pups.

The animal being observed was the mara or Patagonian cavy, a 17.6lb (8kg) hare-like day-active cavy, *Dolichotis patagonum*. (The behavior of the only other member of the subfamily Dolichotinae, the Salt desert cavy, *Dolichotis salinicolum*, is unknown in the wild.) A fundamental element of the mara's social system is the monogamous pair bond. Certainly in captivity, and probably in the wild too, the bond between two animals lasts for life. The drive that impels males to bond with females is so strong that it can lead to "cradle snatching"—adult bachelor males attaching themselves to females while the latter are still infants. Contact between paired animals is maintained primarily by the male who closely follows the female wherever she goes, discouraging approaches from other maras by policing a moving area around her of about 100ft (about 30m) in diameter. In contrast, females appear much less concerned about the whereabouts of their mates. While

foraging, members of a pair maintain contact by means of a low grumble which can be heard only a few feet away.

Monogamy is not common in mammals and in the mara several factors probably combine to favor this social system. Monogamy typically occurs in species where there are opportunities for both parents to care for the young, yet in maras virtually all direct care of the offspring is undertaken by the mother. However, the male does make a considerable indirect investment. Due to the high amount of energy a female uses in bearing and nursing her young she has to spend a far greater proportion of the day feeding than the male—time during which her head is lowered and her vigilance for predators impaired. On the other hand, the male spends a larger proportion of each day scanning and is thus able to warn the female and offspring of danger. Also, by defending the female against the approaches of other maras, he ensures uninterrupted time for her to feed and to care for his young. Furthermore, female maras are sexually receptive only for a few hours twice a year;

▲ **A congregation of maras.** As many as 50–100 maras will associate for a period on the dried-out bed of a shallow lake in the Patagonian thornscrub and at the same time preserve their monogamous pair bonds within the crowd.

▶ **Mother and pups.** A female mara will normally give birth at one time to a maximum of three well-developed (precocial) young. She will nurse them for an hour or more once or twice a day for up to four months.

▶ **A long-legged rodent.** The mara, about 18in (45cm) high, can walk, gallop or run. It has been known to run at speeds up to about 28mph (45km per hour) over long distances.

▼ **Males fight hard** to ward off the challenges of other maras. A male rarely manages to usurp another male's female and even then he will, within a few hours, return to his own mate and allow the deprived male to rejoin his female.

in Patagonia females mate in June or July and then come into heat again in September or October, about 5 hours after giving birth, so a male must stay with his female to ensure he is with her when she is receptive.

Mara pairs generally avoid each other and outside the breeding season it is rare to see pairs within 100ft (about 30m) of each other. Then their home ranges are about 96 acres (40ha). Perhaps the avoidance between pairs is an adaptation to the species' eating habits. Maras feed primarily on short grasses and herbs which are sparsely, but quite evenly, distributed in dry scrub desert. So far, detail of their spatial organization is unknown, but there is at least some overlap in the movements of neighboring pairs. Furthermore, there are some circumstances when, if there is an abundance of food, maras will aggregate. In the Patagonian desert there are shallow lakes, 320ft to several miles in diameter, which contain water for only a few months of the year. When dry, they are sometimes carpeted with short grasses which maras relish. At these times, towards the end of the breeding season (January to March), up to about 100 maras will congregate.

The strikingly cohesive monogamy of maras is noteworthy in its contrast with, and persistence throughout, the breeding season when up to 15 pairs become at least superficially colonial by depositing their young at a communal den. The dens are dug by the females and not subsequently entered by adults. The same den sites are often used for three or more years in a row. Each female gives birth to one to three young at the mouth of the den; the pups soon crawl inside to safety. Although the pups are well developed, moving about and grazing within 24 hours of birth, they remain in the vicinity of the den for up to four months, and are nursed by their mother once or twice a day during this period. Around the den an uneasy truce prevails amongst the pairs whose visits coincide. What social bonds unite them is unknown. The number of pairs of maras breeding at a den varies from 1 to at least 15 and may depend on habitat. Pairs come and go around the den all day and in general at the larger dens at least one pair is always in attendance there.

Even when 20 or more young are kept in a creche, cohabiting amicably, the monogamous bond remains the salient feature of the social system. Each female sniffs the infants seeking to suckle and they respond by proferring their anal regions to the female's nose. Infants clambering to reach one female may differ by at least one month in age. A female's rejection of a usurper can involve a bite and violent shaking. Despite each female's efforts to nurse only her own progeny, interlopers occasionally secure an illicit drink. Although females may thus be engaging in communal nursing they rarely seem to do so as willing collaborators. Indeed, the bombardment of its mother by other youngsters is, to a suckling pup, disadvantageous as the female will interrupt nursing to pursue the pestilent pups.

The reasons why normally unsociable pairs of maras keep their young in a communal creche instead of using separate dens are unknown, but it may be that the larger the number of infants in one place the lower the likelihood that any one will fall victim to a predator; indeed, the more individuals at a den (both adults and young) the more pairs of eyes there are to detect danger. Furthermore, some pairs travel as much as 1.2m (2km) from their home range to the communal den, so the opportunities for shared surveillance of the young may diminish the demands on each pair for protracted attendance at the den. The unusual breeding system of the monogamous mara may thus be a compromise, conferring on the pups benefits derived from coloniality, in an environment wherein association between pairs is otherwise apparently disadvantageous. ABT/DWM

CAPYBARA

Hydrochoerus hydrochaeris
Family: Hydrochoeridae.
Distribution: S America, east of the Andes from Panama to NE Argentina.

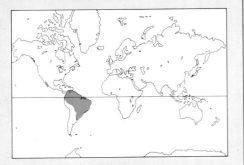

Habitat: open grassland, always near water; also found in a variety of other habitats including tropical rain forest.

Size: head-body length 42–53in (106–134cm), shoulder height 20–24in (50–62cm), weight 77–141lb (35–64kg) for males, 81.6–146lb (37–66kg) for females.

Coat: light brown, consisting of short, abundant hairs in young; adults have long, sparse, bristle-like hairs of variable color, usually brown to reddish.

Gestation: 150 days.
Longevity: 5–10 years.

Subspecies: *H.h. hydrochaeris*, the most widespread, replaced by *H.h. dabbenei* in Paraguay and NW Venezuela and by *H.h. uruguayensis* in Uruguay and E Argentina; *H.h. isthmius* from NW Venezuela, N Colombia and Panama is the smallest of subspecies and is sometimes considered a separate species.

CAPYBARAS are the largest living rodents. They are found only in South America where they live in groups near water. The first European naturalists to visit South America called them Water pigs or Orinoco hogs, though they are neither pigs nor totally aquatic. Although their scientific name, *Hydrochoerus*, means water pig, their nearest relatives are the cavies. Capybaras are the largest living rodents, but the smallest members of their subfamily, Hydrochoerinae. Some extinct forms were twice as long and probably eight times as heavy as modern capybaras, ie they were heavier than the largest modern North American Grizzly bear.

Capybaras are ponderous, barrel-shaped animals. They have no tail and their front legs are shorter than their back legs. Their slightly webbed toes, four in the front feet and three in the back, make them good swimmers, able to stay under water for up to 5 minutes. Their skin is very tough and covered by long, sparse, bristle-like hairs. Their nostrils, eyes and ears are situated near the top of their large, blunt head and hence protrude out of the water when the animal swims. Two pairs of large, typically rodent incisors allow them to eat very short grasses which they grind up with their molar teeth. There are four molars on each side of each lower jaw. The fourth molar is characteristic of the subfamily in being as long as the other three.

They have two kinds of scent glands. One, highly developed in males and almost non-existent in females, is located on top of the snout and called the morrillo (meaning hillock in Spanish). It is a dark, oval-shaped, naked protrusion that produces a copious white, sticky secretion. Both sexes also produce odors from two glandular pockets located on each side of the anus. Male anal glands are filled with easily detachable hairs abundantly coated with layers of hard crystalline calcium salts. Female anal pockets also have hairs but theirs are not detachable and are coated in a greasy secretion rather than with crystalline layers. The proportions of each chemical present in the secretions of individual capybaras are different, potentially providing a means of individual recognition via each personal "olfactory fingerprint."

Capybaras have several distinct vocalizations. Infants and young constantly emit a guttural purr, probably to maintain a contact with their mother or other members of the group. This sound is also produced by losers in aggressive interactions, possibly as an appeasement signal to the opponent. Another vocalization, the alarm bark, is given by the first member of the group to detect a predator. This coughing sound is often repeated several times and the reaction of nearby animals may be to stand alert or to rush into the water.

Capybaras are exclusively herbivorous, feeding mainly on grasses that grow in or near water. They are very efficient grazers and can crop the short dry grasses left at the end of the tropical dry season. Usually they spend the morning resting, then bathe during the hot midday hours; in the late

▲ **The head of the world's largest rodent.**
Because the eyes and ears are small the male's morrillo gland assumes prominence. The smallness of eyes and ears is probably an adaptation for life underwater. The animal's English name is derived from the word used by Guaran-speaking South American Indians. It means "master of the grasses."

◄ **Lazing by a lake.** Capybaras live either in groups averaging 10 in number or in temporary larger aggregations, containing up to 100 individuals, composed of the smaller groups. The situation varies according to season.

afternoon and early evening they graze. At night they alternate rest periods with feeding bouts. Never do they sleep for long periods; rather they doze in short bouts throughout the day.

In the wet season capybaras live in groups of up to 40 animals, but 10 is the average adult group size. Pairs with or without offspring and solitary males are also seen. Solitary males attempt to insinuate themselves into groups, but are rebuffed by the group's males. In the dry season, groups coalesce around the remaining pools to form large temporary aggregations of up to 100 animals. When the wet season returns these large aggregations split up, probably into the original groups that formed them.

Groups of capybaras tend to be closed units where little variation in core membership is observed. A typical group is composed of a dominant male (often recognizable by his large morrillo), one or more females, several infants and young and one or more subordinate males. Among the males there is a hierarchy of dominance, maintained by aggressive interactions consisting mainly of simple chases. Dominant males repeatedly shepherd their subordinates to the periphery of the group but fights are rarely seen. Females are much more tolerant of each other and the details of their social relationships, hierarchical or otherwise, are unknown. Peripheral males may have more fluid affiliation with groups.

Capybaras are found in a wide variety of habitats, ranging from open grasslands to tropical rain forest. Groups may occupy an area varying in size from 4 to 494 acres (2–200ha) with 24.7–49 acres (10–20ha) being most common. Each home range is used mainly, but not exclusively, by one group. Particularly in the dry season, but at other times as well, two or more groups may be seen grazing side by side. Density of capybaras in some areas may be as high as 5 individuals per acre (2 per ha) but lower densities (eg less than 1 per acre) are more frequent.

Capybaras reach sexual maturity at 18 months. In Venezuela and Colombia they appear to breed year round with a marked peak at the beginning of the wet season in May. In Brazil, in more temperate areas, they probably breed just once a year. When a female becomes sexually receptive a male will start a sexual pursuit which may last for an hour or more. The female will walk in and out of the water, repeatedly pausing while the male follows closely behind. The mating will take place in the water. The female stops and the male clambers on her back, sometimes thrusting her under water with his weight. As is usual in rodents, copulation lasts only a few seconds but each sexual pursuit typically involves several mountings.

One hundred and fifty days later up to seven babies are born; four is the average litter size. To give birth the female leaves her

group and walks to nearby cover. Her young are born a few hours later—precocial, able to eat grass within their first week. A few hours after the birth the mother rejoins her group, the young following as soon as they become mobile, three or four days later. Within the group the young appear to suckle indiscriminately from any lactating female; females will nurse young other than their own. The young in a group spend most of their time within a tight-knit creche, moving between nursing females. When they are active they constantly emit a churring purr.

Capybara infants tire quickly and are therefore very vulnerable to predators. They

▲ **Mating** is an aquatic activity, as in hippopotamuses. Unlike the young of the latter, however, capybaras are born on land. For capybaras water is a place of refuge.

► **Scent marking.** ABOVE A male deposits its white sticky secretion from its morrillo.

► **Capybara young.** Capybaras are born after a long gestation of over five months. Even though they emerge in a well-developed (precocial) condition and can eat soon after birth they require over a year before sexual maturity is attained.

Farming Capybara

In Venezuela there has been consumer demand for capybara meat at least since Roman Catholic missionary monks classified it as legitimate lenten fare in the early 16th century, along with terrapins. The similar amphibious habits of these two species presumably misled the monks who supposed that capybaras have an affinity with fish. Today, because of their size, tasty meat, valuable leather and a high reproductive rate capybaras are candidates for both ranching and intensive husbandry.

It has been calculated that where the savannas are irrigated to mollify the effects of the dry season, the optimal capybara population for farming is about 1 animal per acre (about 1.5–3 per ha), which can yield 24lb per acre (27kg per ha) per annum. Those ranches that are licensed to take 30–35 percent of the population at one annual harvest can sustain yields of about 1 capybara per 0.8 acre (or 1 animal per 2ha) in good habitat. An annual cull takes place in February, when reproduction is at a minimum and the animals congregate around

waterholes. Horsemen herd the capybaras which are then surrounded by a cordon of cowboys on foot. An experienced slaughterman then selects adults over 77lb (35kg), excluding pregnant females, and kills them with a blow from a heavy club (see illustration). The average victim weighs 97.4lb (44.2kg) of which 39 percent (38lb; 17.3kg) is dressed meat. These otherwise unmanaged wild populations thus yield annually over 7lb of meat per acre (8kg per ha).

In spite of this yield, farmers feared that large populations of capybaras would compete with domestic stock. In fact because capybaras selectively graze on short vegetation near water they do not compete significantly with cattle (which take more of taller, dry forage) except near wetter, low-lying habitats. There capybaras are actually much more efficient at digesting the plant material than are cattle and horses. So ranching capybara in their natural habitat appears to be, both biologically and economically, a viable adjunct to cattle ranching.

have most to fear from vultures and feral or semiferal dogs which prey almost exclusively on young. Cayman, foxes and other predators may also take young capybaras. Jaguar and smaller cats were certainly important predators in the past, though today they are nearly extinct in most of Venezuela and Colombia. In some areas of Brazil, however, jaguars seize capybaras in substantial numbers.

When a predator approaches a group the first animal to detect it will emit an alarm bark. The normal reaction of other group members is to stand alert but if the danger is very close, or the caller keeps barking, they will all rush into the water where they will form a close aggregation with young in the center and the adults facing outward.

Capybara populations have dropped so substantially in Colombia that c. 1980 the government prohibited capybara hunting. In Venezuela they have been killed since colonial times in areas devoted to cattle ranching. In 1953 the hunting of capybara there became subject to legal regulation and controlled, but to little effect until 1968 when, after a five-year moritorium, a management plan was devised, based on a study of the species biology and ecology. Since then, of the annually censused population in licensed ranches with populations of over 400 capybaras, 30–35 percent are harvested every year. This has apparently resulted in local stabilization of capybara populations. Capybaras are not threatened at present but control on hunting and harvesting must remain if population levels are to be maintained. DWM/EH

OTHER CAVY-LIKE RODENTS

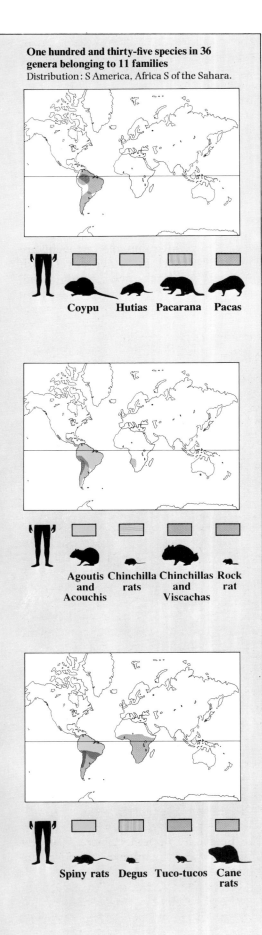

One hundred and thirty-five species in 36 genera belonging to 11 families
Distribution: S America, Africa S of the Sahara.

Coypu Hutias Pacarana Pacas

Agoutis and Acouchis Chinchilla rats Chinchillas and Viscachas Rock rat

Spiny rats Degus Tuco-tucos Cane rats

THE families assembled in this chapter are a disparate assemblage of ancient rodent fauna of South America, a few of which have migrated into Central and North America since the two continents were rejoined in the Pliocene era (7–3 million years ago). The diversity they show is in marked contrast to the relative homogeneity of the more recent rodents.

In this group there are small rodents and large rodents; some are covered with barbed spines, others have soft silky fur. Some species are common and widespread, others known only from a few museum specimens. They inhabit forests and grasslands, water and rocky deserts, coastal plains and high mountains; some are solitary, others colonial. Many species are eaten by humans, others prized for their fur; some are pests, others carry the diseases of man and domestic animals.

The larger species, such as agoutis, paca, pacarana and viscacha, are prey for the large and medium-sized species of carnivore (jaguar, ocelot, pampas cat, maned wolf, bush dog, foxes etc). They are herbivorous and may be considered as the South American ecological equivalent of the vast array of ungulate herbivores which are so important in the African ecosystems. It is thought that these rodents radiated into this role as the primitive native herbivores became extinct and before the arrival of the new fauna from the north.

The **coypu** is a large robust rodent, weighing up to 22lb (10kg). It lives in burrows in river banks, feeds on water plants and is an expert swimmer. Families of up to 10 young are recorded—most of the female's mammae are situated in a row high on the side to enable her to feed the precocial young while swimming.

The coypu has small rounded ears and webbed hind feet. Its fur is darkish brown-yellow and the tip of its muzzle white. Its outer hair is long and coarse, covering the thick soft underfur known in the fur trade as nutria—a word corrupted from the Spanish word for otter.

Hutias are found only in the West Indies, living in forests and plantations and eating not only vegetation but sometimes small animals such as lizards. Hutias are robust, short-legged rats ranging in length from 8 to about 24in (20–60cm) and weighing up to 15lb (7kg). Their fur is rough but with a soft underfur. Two species of the genus *Plagiodontia* are known only from subfossil remains found in caves and kitchen middens. Three other genera, *Hexalobodon*, *Aphaetreus* and *Isolobodon*, are known from similar subfossil bones and are thought to have become extinct within historical time. Several other species have recently become extinct thanks to human destructiveness. Among living representatives of the group, *Capromys* is a forest dweller weighing up to 15lb (7kg) and is hunted for its flesh in Cuba, and *Geocapromys* is a short-tailed nocturnal form which lives on leaves, bark and twigs and weighs up to 4.4lb (2kg). Little is known of the biology of these animals and several species are thought to be in peril from the Burmese mongoose which has been introduced to the West Indies.

The **pacarana** is a slow-moving, robust animal, weighing up to 22–33lb (10–15kg), which resembles a spineless porcupine. It is coarse haired, black or brown with two more or less continuous white stripes on each side. It has broad, heavily clawed feet and uses its forepaws to hold food while eating. Its tail is about one-quarter of its head-body length which may reach 31in (80cm). A forest-dwelling species, seldom encountered, about which little is known,

◀ **Coypu afloat.** Coypus are probably the most aquatic of cavy-like rodents. They have webbed feet, spend most of their waking hours in water and live in burrows bored in river banks.

▼ **Coypu abroad.** Coypu fur, labeled as "nutria" fur, has been popular since the early 19th century. Consequently coypu farms have been established on every continent except Antarctica, but in many countries animals have become feral, including this inhabitant of East Africa. Some have also become pests.

this inoffensive herbivore is prey for jaguar, ocelot and other medium-sized carnivores and is hunted for food by man. It has a remarkably long gestation period, which can vary from 220 to 280 days.

About 10 species of **agouti** and two of **acouchi** make up the family Dasyproctidae, all large rodents sometimes considered to belong to the same family as the pacas. They are relatively common animals although secretive: they hide in burrows and become nocturnal in areas where they are disturbed. The coat of agoutis is orange to brown or blackish above, yellowish to white below, with a contrasting rump color. Acouchis are reddish to blackish green above, yellowish below, with a bright color on the head. Agoutis live mainly on fallen fruits. They are attracted to the sound of ripe fruits hitting the ground. When food is abundant they carefully bury some for use in time of scarcity. This behavior is important in dispersing the seeds of many species of forest trees. Acouchis, which have relatively long tails, are rarely seen and their biology in the wild is almost unknown. These animals are hunted for food and preyed upon by a variety of carnivores.

Chinchilla rats are soft-furred rats with short tails. *Abrocoma cinerea* is smallest, with a head-body length of 6–8in (15–20cm) and tail length 2.5–6in (6–15cm), with comparable figures for *A. bennetti*, 8–10in (20–25cm) and 5–7in (13–18cm), respectively. Both have soft dense underfur, overlain with long, fine guardhairs; coat color is silver gray or brown above, white or brown below. Their pelts are occasionally sold but are of much poorer quality than the pelts of true chinchillas. They live in tunnels and crevices in colonies from sea level to the high Andes of southwestern South America. *Abrocoma bennetti* is distinguished by having more ribs than any other rodent (17 pairs).

Spiny rats comprise some 15 genera and about 56 species and are a peculiar assemblage of robust, medium-sized, herbivorous rodents, most of which have a spiny or bristly coat. (Some species however are soft furred.) Some are very common and widespread while others are extremely rare; some have short tails, others long tails (both types having a tendency to become lost), and some are arboreal, others burrowing. The taxonomy of some genera is poorly understood and the numbers of species are tentative.

The body form in this family is correlated to life style. Robust, short-tailed forms (eg *Clyomys, Carterodon* and *Euryzygomatomys*) are burrowing savanna species; relatively slender, long-tailed forms are arboreal (eg *Thrinacodus, Dactylomys, Kannabateomys*). Of the intermediate forms *Proechimys, Isothrix, Hoplomys* and *Cercomys* are more or less terrestrial and *Mesomys, Louchothrix, Echimys* and *Diplomys* are mostly arboreal. Of the many genera only two have produced any number of species: *Proechimys* and *Echimys*.

Chinchillas and **viscachas** are soft-furred animals with long, strong hind legs, large ears and tails up to one-third the length of the body. Their teeth are characteristically divided into transverse plates. The Plains viscacha inhabits pampas and shrub in Argentina. It has soft underfur overlain with stiffish guardhairs and is brown or gray above, white below, with a distinctive black and white face. Plains viscachas are sexually dimorphic with males at 18lb (8kg) almost twice the size of females. Mountain viscachas occur in the Andes of Peru, Bolivia, Argentina and Chile and have thick soft fur, gray or brown above, whitish or grayish below. Chinchillas, which weigh only up to 1.8lb (0.8kg), inhabit the same regions as mountain viscachas. They have very dense fur, bluish gray above, yellowish below. All species are subject to pressure from human hunting—chinchillas have been hunted for their valuable fur to near extinction, mountain viscachas are prized for both food and fur and the Plains viscachas compete for grazing with domestic animals, ten viscachas eating as much as a

sheep. In addition they destroy pasture with their acidic urine, and so undermine the pampas that men, horses and cattle are often injured by falling into their concealed tunnels. The animals collect a variety of materials (bones, sticks and stones) lying loose in their surroundings and heap them in piles above the entrances to their burrows.

Degus (or **octodonts**) occur in southern South America from sea level to about 10,000ft (3,000m). The name octodonts refers to the worn enamel surface of their teeth which forms a pattern in the shape of a figure eight. All are robust rodents, with a head-body length of 5–8in (12–20cm) and tail length of 1.6–7in (4–18cm), and a long silky coat. Degus and choz choz are gray to brown above, creamy yellow or white below. White corucos are brown or black and Rock rats dark brown all over. Viscacha rats are buffy above, whitish below, with a particularly bushy tail up to 7in (18cm) long. Most are adapted to digging, particularly the Rock rats, or to living in rock crevices. Corucos have broad well-developed incisors, used in burrowing. They eat bulbs, tubers, bark and cactus.

Tuco-tucos comprise one genus with about 33 species, which inhabit sandy soils from the altoplano of Peru to Tierra del fuego in a wide variety of vegetation types. They are fossorial (adapted for digging), building extensive burrow systems and may be compared to the pocket gophers of North

The 12 families of other Cavy-like rodents

Coypu
Family: Myocastoridae.
1 species, *Myocastor coypus*, S Brazil, Paraguay, Uruguay, Bolivia, Argentina, Chile; feral populations in N America, N Asia, E Africa, Europe.

Hutias
Family: Capromyidae.
12 species in 4 genera.
W Indies. Genera: **Cuban hutias** (*Capromys*), 9 species; **Bahaman and Jamaican hutias** (*Geocapromys*), 2 species; *Isolobodon*, 3 species, recently extinct; **Hispaniolan hutias** (*Plagiodontia*), 1 living and 1 recently extinct species.

Pacarana
Family: Dinomyidae.
1 species, *Dinomys branicki*, Colombia, Ecuador, Peru, Brazil, Bolivia, lower slopes of Andes.

Pacas
Family: Agoutidae.
2 species in 1 genus.
Mexico S to S Brazil.

Agoutis and Acouchis
Family: Dasyproctidae.

13 species in 2 genera.
Mexico S to S Brazil. Genera: **agoutis** (*Dasyprocta*), 11 species; **acouchis** (*Myoprocta*), 2 species.

Chinchilla rats
Family: Abrocomidae.
2 species in 1 genus.
Altoplano Peru, SW Bolivia, Chile, NW Argentina.
Species: *Abrocoma bennetti, A. cinerea.*

Spiny rats
Family: Echimyidae.
56 species in 15 genera.
Nicaragua S to Peru, Bolivia, Paraguay, S Brazil.
Genera: **spiny rats** (*Proechimys*), 22 species; **spiny rats** (*Echimys*), 10 species; **Guiara** (*Euryzygomatomys spinosus*); **Owl's rat** (*Carterodon sulcidens*); **Thickspined rat** (*Hoplomys gymnurus*); *Clyomys laticeps*; **Rabudos** (*Cercomys cunicularis*); *Mesomys*, 4 species; *Lonchothrix emiliae*; **toros** (*Isothrix*), 3 species; **soft-furred spiny rats** (*Diplomys*), 3 species; **corocoro** (*Dactylomys*), 3 species; *Kannabateomys amblyonyx*; Thrinacodus, 2 species.

Chinchillas and Viscachas
Family: Chinchillidae.
6 species in 3 genera.

Peru, Argentina, Chile, Bolivia. Genera: **Plains viscacha** (*Lagostomus maximus*); **mountain viscachas** (*Lagidium*), 3 species; **chinchillas** (*Chinchilla*), 2 species.

Degus or **Octodonts**
Family: Octodontidae.
8 species in 5 genera.
Peru, Bolivia, Argentina, Chile, mainly in Andes (some along coast). Genera: **Rock rat** (*Aconaemys fuscus*); **degus** (*Octodon*), 3 species; **Choz choz** (*Octodontomys gliroides*); **viscacha rats** (*Octomys*), 2 species; **Coruros** (*Spalacopus cyanus*).

Tuco-tucos
Family: Ctenomyidae.
33 species in 1 genus (*Ctenomys*).
Peru S to Tierra del Fuego.

Cane rats
Family: Thryonomyidae.
2 species in 1 genus (*Thryonomys*).
Africa S of the Sahara.

African rock rat
Family: Petromyidae.
1 species, *Petromus typicus*, W South Africa N to SW Angola.

▲ **Representatives of 10 families** of other cavy-like rodents. (**1**) A chinchilla (family Chinchillidae). (**2**) An American spiny rat climbing (family Echimyidae). (**3**) A hutia sunning itself (family Capromyidae). (**4**) A cane rat (family Thryonomyidae). (**5**) A Rock rat (family Petromyidae). (**6**) A pacarana (*Dinomys branickii*) feeding. (**7**) A chinchilla rat (*Abrocoma bennetti*). (**8**) A paca (*Agouti paca*). (**9**) A tucotuco (*Ctenomys opimus*) excavating with its incisors. (**10**) A degu (*Octodon degus*).

stored in cells in the tunnels but such stores are often left to decay.

Cane rats and the African **rock rat** are African rodents sufficiently distinctive to be placed in families of their own. They appear to be related to the African porcupines which in turn may have some affinities with the cavy-like rodents of the Neotropics.

There are probably only two species of cane rat although many varieties have been described. They are robust rats weighing up to 20lb (9kg), occasionally more. Head and body length ranges from 14 to 24in (35–60cm), tail length from 3 to 10in (7–25cm).

Cane rats prefer to live near water and eat a variety of vegetation, especially grasses, and as their name suggests they can be pests of plantations. They are prey for leopard, mongoose and python in addition to being hunted for food by man in many parts of their range. Their coat is coarse and bristly, the bristles being flattened and grooved along their length. There is no underfur. Coat color is brown speckled with yellow or gray above, buffy white below. Preferred habitats are reed beds, marshes and the margins of lakes and rivers. In some parts of their range cane rats breed all year round but most young are born between June and August with two to four in a litter. Of the two species *Thryonomys swinderianus* is the larger, attaining weights of up to 20lb (9kg) and head and body length of up to 24in (60cm); *T. gregorianus* may occasionally reach 15lb (7kg) and head and body length of 20in (50cm). The latter is said to be less aquatic than *T. swinderianus*.

The single species that belongs to the family Petromyidae is known as the African rock rat or dassie. It has a flattened skull and very flexible ribs which allow it to squeeze into narrow rock crevices. Its color is very variable and mimics the color of the rocks amongst which the animal lives. It is particularly active at dawn and dusk, and lives on vegetable matter such as leaves, berries and seeds. Head and body length varies from 5.5in to 8in (14–20cm), tail length from 5in to 7in (13–18cm). Its coat is long and soft but there is no recognizable underfur. Shades of gray, yellow and buff predominate. Its preferred habitat is rocky hillsides of southwestern South Africa and the animal is found only where there are rocks for shelter. Its tail has hairs scattered throughout its length and long white hairs at the tip. Rock rats breed only once per year and produce two young to a litter, and thus have a relatively slow reproduction rate for small mammals in the tropics.   IRB

America. Their hind feet bear strong bristle fringes, their ears are small, their claws strong and their incisors well developed. Their size ranges from head-body length 6–10in (15–25cm), tail length 2.4–4.3in (6–11cm) and they weigh up to 1.6lb (0.7kg). Their coat is very variable, ranging from gray or buff to brown and reddish brown above, lighter below; their hairs can be long or short, usually dense, but never bristly. The name "tuco-tucos" refers to their calls.

Their shallow burrows may have several entrances and by opening and plugging tunnels as necessary *Ctenomys* can regulate the burrow temperature, which is normally maintained at about 68–72°F (20–22°C). Feeding is often accomplished by pulling roots down into a tunnel. Food is often

OLD WORLD PORCUPINES

Family: Hystricidae
Eleven species in 4 genera and 2 subfamilies.
Distribution: Africa, Asia.

Habitat: varies from dense forest to semidesert.

Size: ranges from head-body length of 14.6–18.5in (37–47cm) and weight 3.3–7.7lb (1.5–3.5kg) in the brush-tailed procupines to head-body length 23.6–32.7in (60–83cm) and weight 28.6–59.4lb (13–27kg) in the crested porcupines.

Gestation: 90 days for the Indian porcupine, 93–4 days for the Cape porcupine, 100–10 days for the African brush-tailed porcupine, 105 days for the Himalayan porcupine, 112 days for the African porcupine.

Longevity: approximately 21 years recorded for the crested porcupines (in captivity).

Brush-tailed porcupines
Genus *Atherurus.*
C Africa and Asia; forests; brown to dark brown bristles cover most of the body; some single color quills on the back. Two species: **African brush-tailed porcupine** (*A. africanus*); **Asiatic brush-tailed porcupine** (*A. macrourus*).

Indonesian porcupines
Genus *Thecurus.*
Coat: dark brown in front, black on posterior; body densely covered with flattened flexible spines; quills have a white base and tip with central parts black; rattling quills on the tail are hollow. Three species: **Bornean porcupine** (*T. crassispinis*); **Philippine porcupine** (*T. pumilis*); **Sumatran porcupine** (*T. sumatrae*).

Bornean long-tailed porcupine
Trichys lipura.
Body covered with brownish flexible bristles; head and underparts hairy.

Crested porcupines
Genus *Hystrix.*
Africa, India, SE Asia, Sumatra, Java and neighboring islands, S Europe; recently introduced to Great Britain; hair on back of long, stout, cylindrical black and white erectile spines and quills; body covered with black bristles; grayish crest well developed. Five species: **African porcupine** (*H. cristata*); **Cape porcupine** (*H. africaeaustralis*); **Himalayan porcupine** (*H. hodgsoni*); **Indian porcupine** (*H. indica*); **Malayan porcupine** (*H. brachyura*).

Porcupines have a peculiar appearance, due to parts of their bodies being covered with quills and spines. It is often thought that they are related to hedgehogs or pigs. Their closest relatives, however, are guinea pigs, chinchillas, capybaras, agoutis, viscachas and cane rats. Many of these have in common an extraordinary appearance, and are well known, at least to zoologists, for their unusual ways of solving problems involved in reproduction.

The Old World porcupines belong to two distinct subfamilies: Atherurinae and Hystricinae. The brush-tailed porcupines (of the subfamily Atherurinae) have long, slender tails which end in a tuft of white, long, stiff hairs. In the genus *Atherurus* these hairs have hollow expanded sections at intervals which produce a rustling sound when the tail is shaken. Their elongated bodies and short legs are covered with short, flat, chocolate-colored, sharp bristles, with only a few long quills on the back. Crested porcupines (subfamily Hystricinae), on the other hand, have short tails surrounded by an array of cylindrical, stout, sharp quills with the tip of the tail being armed with a cluster of hollow open-ended quills with a narrow stalk-like base, which produce a characteristic rattling sound when the tail is shaken, serving as a warning signal when the animal is annoyed by an intruder. The posterior two-thirds of the upper parts and flanks of the body are covered with black and white spines, up to 20in (50cm) in length, and sharp cylindrical quills which cannot be projected at the enemy but which are on contact easily detached.

When aggressive, porcupines erect their spines and quills, stamp their hind feet, rattle their quills and make a grunting noise. If threatened further they will turn their rump towards an intruder and defend themselves further by quickly running sideways or backwards into the enemy. If the quills penetrate the skin of the enemy they become stuck and detach. Although not poisonous, embedded quills often cause septic wounds which can prove fatal for their natural predators, eg lions and leopards.

The remainder of the stout body is covered with flat, coarse, black bristles. Most species have a crest that can be erected, extending from the top of the head to the shoulders, consisting of long, coarse hair. Their heads are blunt and exceptionally broad across the nostrils. Their small pig-like eyes are set far back on the sides of the head. The sexes look alike. The female's mammary glands are situated on the side of the body.

Most Old World porcupines are vegetarians. In their natural habitats they feed on the roots, bulbs, fruit and berries of a variety of plants. When they enter cultivated areas they will eat such crops as groundnuts, potatoes, pumpkins, melons and maize. African porcupines are reported as able to feed on plant species known to be poisonous to cattle. Porcupines manipulate their food with their front feet and while eating hold it against the ground. Their chisel-like incisors enable them to gnaw effectively. Porcupines are solitary feeders but groups of two or three individuals, comprising one or two adults and offspring, have been observed on occasion. Porcupine shelters often contain accumulations of bones, carried in for gnawing, either for sharpening teeth or as a source of phosphates. Brush-tail porcupines are active tree-climbers and feed on a variety of fruits.

Detailed information on the reproductive biology of porcupines is only available for the Cape porcupine. Sexual behavior leading to copulation is normally initiated by a female who, after having approached a male, or after being approached by a male, will take up the sexual posture, in which she raises her rump and tail and holds the rest of her body close to the surface. The male mounts the female by standing bipedally behind the female with his forepaws resting on her back. Thrusting only occurs after intromission, which only occurs when the female is in heat (every 28–36 days), when the vaginal closure membrane becomes perforated. Sexual behavior without intromission is exhibited during all stages of the sexual cycle.

The young are born in grass-lined chambers which form part of an underground burrow system. At birth they are unusually precocial: fully furred, eyes open, with bristles and heralds of future quills already present. These harden within a few hours after birth. They weigh 10.6–11.6oz (300–330g) and start to nibble on solids between nine and 14 days. They begin to feed at four to six weeks but are nursed for 13–19 weeks, when they weigh 7.7–10.4lb (3.5–4.7kg). Litter size varies but 60 percent of all births produce one young and 30 percent produce twins. Three is normally the maximum number. Females produce only one litter per year and although not seasonal when kept in captivity they breed only in summer in their natural habitats. In a colony of porcupines all members protect the young, adult males playing an important role by being particularly aggressive towards invaders. Sexual maturity is at-

▲ On a desert dune, a crested porcupine. These porcupines have amazing versatility, in some ways resembling that of some mouse-like rodents. They can live in deserts, steppe, rocky areas and forest. They shelter in existing holes and crevices or dig their own burrows. They are also successful in dealing with predators. Such factors probably enable them to live long, often for 12–15 years.

▼ Amidst grass in Malaysia: an Asiatic brush-tailed porcupine. This is a nocturnal species which lives in groups in forests.

tained at an age of two years.

All evidence suggests that Cape porcupines live in colonies comprising at least an adult pair and their consecutive litters, with as many as six to eight individuals occupying a burrow system. Females as well as males will be aggressive towards strange males and females, irrespective of the sexual state of the female. Both will also accompany young up to the age of six or seven months when they go out foraging.

Population density in the semiarid regions of South Africa varies from 2.5 to 75 individuals per sq mi (29 per sq km) with approximately 40 percent of populations comprising individuals less than one year old.

No information is available about territoriality or the size of home ranges but porcupines have been reported to forage up to 10mi (16km) from their burrows per night. They move along well-defined tracks, almost exclusively by night. Porcupines are catholic in their habitat requirements, providing they have shelter to lie in during the day. They take refuge in crevices in rocks, in

caves or in abandoned aardvark holes in the ground, which they modify by further digging to suite their own requirements.

Porcupines are often reported as a menace to crop-producing farmers. They are destroyed by various methods. Their flesh is enjoyed by indigenous people throughout Africa and the killing of porcupines, whenever the opportunity arises, has apparently become a favored pastime. They still, however, occur in great numbers throughout Africa, thanks to the near absence of their natural predators (lions, leopards, hyenas) over much of their range and also because of the increase in food production through the cultivation of agricultural crops. African porcupines also carry fleas, which are responsible for the spread of bubonic plague, and ticks, which spread babesiasis, rickettsiasis and theilerioses. Brush-tailed porcupines are known to be hosts of the malarial parasite *Plasmodium atheruri*. In spite of being killed in large numbers there is no reason to believe that porcupines are endangered.   RJvA

GUNDIS

Family: Ctenodactylidae
Five species in 4 genera.
Distribution: N Africa.

Habitat: rock outcrops in deserts.

Size: ranges from head-body length 6.8–7.2in (17–18cm), tail length 1.1–1.3in (2.8–3.2cm), weight 6.3–6.9oz (178–195g) in the Felou gundi to head-body length 6.9–7.1in (17.2–17.8cm), tail length 2–2.2in (5.2–5.6cm), weight 6.2–6.3oz (175–180g) in Speke's gundi.

Gestation: 56 days in the Desert or Sahara gundi (unknown for other species).

Longevity: 3–4 years (10 years recorded for Speke's gundi in captivity).

North African gundi
Ctenodactylus gundi
SE Morocco, N Algeria, Tunisia, Libya, occurring in arid rock outcrops.

Desert gundi
Ctenodactylus vali
Desert gundi or Sahara gundi.
SE Morocco, NW Algeria, Libya, occurring in desert rock outcrops.

Mzab gundi
Massoutiera mzabi
Mzab gundi or Lataste's gundi.
Algeria, Niger, Chad, occurring in desert and mountain rock outcrops.

Felou gundi
Felovia vae
SW Mali, Mauritania, occurring in arid and semiarid rock outcrops.

Speke's gundi
Pectinator spekei
Speke's gundi or East African gundi.
Ethiopia, Somalia, N Kenya occurring in arid and semiarid rock outcrops.

▶ **Mzab gundi.** ABOVE An extraordinary feature of this gundi is that its ears are flat and immovable.

▶ **Speke's gundi,** from East Africa, has a range of well-developed vocalizations.

THE first gundi was found in Tripoli in 1774 and called the gundi-mouse (gundi is the North African name). In the mid-19th century the explorer John Speke shot gundis in the coastal hills of Somalia, and later French naturalists found three more species; skins and skulls began to arrive in museums. But no attempt was made to study the ecology of the animal. Some authors said gundis were nocturnal; others, diurnal. Some said they dug burrows, others said they did not. Some said they made nests. Some heard them whistling; others, chirping like birds—and there were fantastic tales about them combing themselves in the moonlight with their hind feet. The family name is Ctenodactylidae which means comb-toes. In 1908 two French doctors isolated a protozoan parasite (now known to occur in almost every mammal) from the spleen of a North African gundi and called it *Toxoplasma gondii*.

Gundis have short legs, short tails, flat ears, big eyes and long whiskers. Crouched on a rock in the sun with the wind blowing through their soft fur they look like powder puffs.

The North African and the Desert gundi have tiny wispy tails but the other three have fans which they use as balancers. Speke's gundi has the largest and most elaborate fan which it uses in social displays. Gundis also have rows of stiff bristles on the two inner toes of each hind foot which stand out white against the dark claws. They use the combs for scratching. Sharp claws adapted to gripping rocks would destroy the soft fur coat that insulates them from extremes of heat and cold. The rapid circular scratch of the rump with the combed instep is characteristic of gundis.

The gundi's big eyes convinced some authors that the animal was nocturnal. In fact the gundi is adapted to popping out of sunlight into dark rock shelters. Equally the gundi can flatten its ribs to squeeze into a crack in the rocks.

Gundis are herbivores: they eat the leaves, stalks, flowers and seeds of almost any desert plant, including grass and acacia. Their incisors lack the hard orange enamel that is typical of most rodents. Gundis are not, therefore, great gnawers. Food is scarce in the desert and gundis must forage over long distances—sometimes as much as 0.6mi (1km) a morning. Regular foraging is essential because gundis do not store food. Home range size varies from a few square feet to 1.9sq mi (3sq km).

Foraging over long distances generates body heat which can be dangerous on a hot

desert day. It is unusual for small desert mammals to be active in daytime but gundis behave a bit like lizards. In the early morning they sunbathe until the temperature rises above 68°F (20°C) and then forage for food. After a quick feed they flatten themselves again on the warm rocks. Thus they make use of the sun to keep their bodies warm and to speed digestion. It is an economical way of making the most of scarce food. By the time the temperature has reached 90°F (32°C), the gundis have taken shelter from the sun under the rocks and will not come out again until the temperature drops in the afternoon. When long foraging expeditions are necessary gundis alternate feeding in the sun and cooling off in the shade. In extreme drought gundis eat at dawn when plants contain most moisture. They obtain all the water they need from plants; their kidneys have long tubules for absorbing water. Their urine can be concentrated if plants dry out completely.

But this emergency response can only be sustained for a limited period.

Gundis are gregarious, living in colonies that vary in density from the Mzab gundi's 0.12 per acre (0.3 per ha) to over 40 per acre (100 per ha) for Speke's gundi. Density is related to the food supply and the terrain. Within colonies there are family territories occupied by a male and female and juveniles or by several females and offspring. Gundis do not make nests and the "home shelter" is often temporary. Characteristically a shelter retains the day's heat through a cold night and provides cool draughts on a hot day. In winter, gundis pile on top of one another for warmth, with juveniles shielded from the crush by their mother or draped in the soft fur at the back of her neck.

Each species of gundi has its own repertoire of sounds, varying from the infrequent chirp of the Mzab gundi to the complex chirps and chuckles and whistles of Speke's gundi. In the dry desert air and the rocky terrain their low-pitched alert calls carry well. Short sharp calls warn of predatory birds; gundis within range will disappear under the rocks. Longer calls signify ground predators and inform the predator it has been spotted. The Felou gundi's harsh chee-chee will continue as long as the predator prowls around. Long complex chirps and whistles can be a form of greeting or recognition. The *Ctenodactylus* species—whose ranges overlap—produce the most different sounds: the North African gundi chirps, the Desert gundi whistles. Thus members can recognize their own species. All gundis thump with their hind feet when alarmed. Their flat ears give good all-round hearing and a smooth outline for maneuvering among rocks. The bony ear capsules of the skull are huge, like those of many other desert rodents. The acute hearing is important for picking up weak low-frequency sounds of predators—sliding snake or flapping hawk—and for finding parked young. Right from the start, young are left in rock shelters while the mother forages. The young are born fully furred and open-eyed. The noise they set up—a continuous chirruping—helps the mother to home in on the temporary shelter.

The young have few opportunities to suckle: from the mother's first foraging expedition onwards they are weaned on chewed leaves. (They are fully weaned after about 4 weeks.) The mother has four nipples—the average litter size is two—two on her flanks and two on her chest. But a gundi has little milk to spare in the dry heat of the desert.   WG

AFRICAN MOLE-RATS

Family: Bathyergidae
Nine species in 5 genera.
Distribution: Africa S of the Sahara.

Habitat: underground in different types of soil and sand.

Size: ranges from head-body length 3.5–4.7in (9–12cm), weight 1–2.1oz (30–60g) in the Naked mole-rat to head-body length 11.8in (30cm), weight 26–63oz (750–1,800g) in the genus *Bathyergus*.

Gestation: 70 days in the Naked mole-rat; unknown for all other genera.

Longevity: unknown (captive Naked mole-rats have lived for over 10 years).

Subfamily Bathyerginae

Dune mole-rats
Genus *Bathyergus*.
Sandy coastal soils of S Africa; the largest mole-rats. Two species including: **Cape dune mole-rat** (*B. suillus*).

Subfamily Georychinae

Common mole-rats
Genus *Cryptomys*.
W, C and S Africa; the most widespread genus. Three species including: **Common mole-rat** (*C. hottentotus*).

Cape mole-rat
Georychus capensis.
Cape Province of the Republic of S Africa, along the coast from the SW to the E.

Silvery mole-rats
Genus *Heliophobius*.
C and E Africa. Two species including: **Silvery mole-rate** (*H. argenteocinereus*).

Naked mole-rat
Heterocephalus glaber.
Arid regions of Ethiopia, Somalia and Kenya.

▶ **Shaped like a cylinder,** ABOVE, the Common mole-rat. Members of its genus may be the inhabitants of the longest constructed and maintained burrows of any animal (up to 1,150ft, 350m).

▶ **Rodent digging power,** the head of a Cape mole-rat.

THE mole-rat is a rat-like rodent that has become totally conditioned to life underground and has assumed a mole-like existence. Consequently its anatomy and life-style are distinctive if not unusual: it digs out an extensive system of semipermanent burrows for foraging—to which are attached sleeping areas and rooms for storing food—and throws up the soil it excavates as "mole hills." Most rodents of comparable size grow rapidly and live for only a couple of years. Naked mole-rats, however, take over a year to attain adult size and can live for several years. All these features must have contributed to the evolution of their highly structured social system.

Many of the physical features of the mole-rat are designed for its life underground where boring through soil and pushing it up as molehills requires considerable power and energy. Efficiency in effort is essential. Mole-rats have cylindrical bodies with short limbs so as to fit as compactly as possible within the diameter of a burrow. Their loose skin helps them to turn within a confined space: a mole-rat can almost somersault within its skin as it turns. Mole-rats can also move rapidly backwards with ease and so when moving in a burrow tend to shunt to and fro without turning round.

All genera, except dune mole-rats, use chisel-like incisors protruding out of the mouth cavity for digging. To prevent soil from entering the mouth while digging there are well-haired lip-folds behind the incisors. Dune mole-rats dig with the long

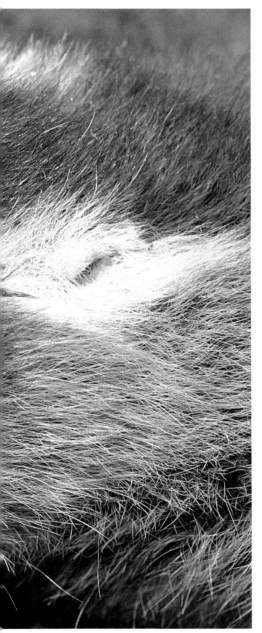

senses of smell and hearing are well developed and their noses and ears are modified on the outside so as to cope with the problems of living in a sandy environment: the nostrils can be closed during digging while the protruding parts of the ears (pinnae) have been lost.

Mole-rats are vegetarians and obtain their food by digging foraging tunnels. These enable them to find and collect roots, storage organs (geophytes) and even the aerial portions of plants without having to come above ground. The large Dune mole-rat, less well equipped, cannot afford to be a specialist feeder and lives mainly on whatever it encounters—grass, herbs and geophytes. On the other hand, the small Common and Naked mole-rats are more selective and live entirely on geophytes and roots. This may be possible because they displace less soil as they dig and, being social, share the cost of digging and finding food between members of the colony. Foraging burrows can be very extensive: one system containing 10 adults and 3 young Common mole-rats ran for 0.6mi (1km). Burrow length depends on the number, ages and conditions of mole-rats in a system and on the abundance of food items. Apart from providing nest, food storage and toilet areas, the entire burrow system is dug in search of food.

Burrowing activity normally increases just after the rainy season, when the soil is soft, moist and easily worked. It appears that food sources located at this time are exploited at a later date. Small food items are stored (by Common and Cape mole-rats for example), larger items are left growing *in situ* and are gradually hollowed out by the mole-rats, thus ensuring a constantly fresh and growing food supply. In some areas where Naked mole-rats occur, tubers may weigh as much as 110lb (50kg). When feeding, the mole-rat holds small items with its forefeet, shakes them free of soil, cuts them up into small pieces with its incisors and then chews these with its cheekteeth.

In southwestern Cape Province, South Africa, differences in diet, burrow diameter and depth, and perhaps social organization, enable three genera to occupy different niches within the same geographical area—indeed in the same field: *Bathyergus*, *Georychus*, and *Cryptomys*. This sympatry is unusual for burrowing mammals, where the normal pattern is for one species to occupy one area exclusively.

The social behavior of three genera (*Bathyergus*, *Georychus* and *Heliophobius*) follows the normal pattern for subterranean mammals: they are solitary. Nothing is

claws on their forefeet and so are less efficient at mining. Moreover their body size is larger. These disadvantages seem to restrict them to areas with easily dug sandy soil. This difference in digging method is reflected in the teeth of mole-rats: the incisors of tooth-diggers have roots that extend back behind the row of cheekteeth.

When a mole-rat is tunneling it pulls the soil under its body with its forefeet. Then, with the body weight supported by the forefeet, both hind feet are brought forward, collect the soil and kick it behind the animal. Once a pile of soil has accumulated the mole-rat reverses along the burrow with the soil and pushes it up a side branch to the surface. In all but the Naked mole-rats solid cores of soil are forced out onto the surface to form a molehill. Naked mole-rats kick a fine spray of soil out of an open hole—an "active" hole looks like an erupting volcano. A number of Naked mole-rats cooperate in digging, one animal excavating, a number transporting soil and another individual kicking it out of the hole. In all mole-rats the hind feet are fringed by stiff hairs which help hold the soil down during digging movements.

Because mole-rats live in complete darkness for most of their lives their eyes are small. (Interestingly, the Cape mole-rat, which spends some time on the surface, has eyes larger than those of other mole-rats.) Evidence derived from experiments suggests that the eyes are sightless: only their surface is used, to detect air currents which would indicate damage to the burrow system. Indeed if damage occurs they rapidly repair it. Touch is important in finding their way round the burrow system; many genera have long touch-sensitive hairs scattered over their bodies (in the Naked mole-rat these are the only remaining hairs). Their

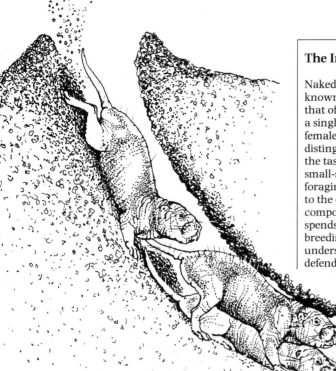

The Insect-like Rodent

Naked mole-rats are the only mammals known to have a colony structure similar to that of social insects. Within each colony only a single pair breeds; the remaining males and females belong to castes which are distinguishable by differences in size and by the tasks they perform. More numerous, small-sized, working-caste mole-rats dig the foraging burrows and carry food and nesting to the communal nest. A nonworking caste, composed of fewer, large-sized individuals, spends most of the time in the nest with the breeding female: its role is not yet clearly understood but it may be involved in defending the colony.

energies of colony members along specific avenues (some finding food, others bearing young) these mole-rats can survive in areas where single or pairs of mole-rats cannot.

Experiments have shown that nonbreeding members of Naked mole-rat colonies are not sterile. They can found new colonies and can also replace the breeding animals if they die. In this latter case, if the colony is an established one and otherwise undisturbed several females initially show signs of sexual activity, but one will grow rapidly and become sexually dominant within a few weeks of the death of the former breeding female. Usually no fighting occurs, suggesting that a hierarchy exists within the colony and that the successor is already determined.

If, therefore, the nonbreeding mole-rats are

known of mating behavior but their young disperse soon after weaning. In *Georychus* the dispersing young appear to burrow away from the parent system and block up the linking burrows; this probably also occurs in the other solitary genera and would ensure that the young are protected from predators during this otherwise very vulnerable phase in their life history. If forcibly kept together in captivity, levels of aggression build up until litter mates will kill each other. The Common mole-rat occurs in pairs or small groups about which little is known, while Naked mole-rat colonies may contain over 80 individuals, but here only a pair of mole-rats breed (see box).

Because they live in a well-protected, safe environment, mole-rats, unlike most rodents, have little to fear in the way of predators. Perhaps as a consequence their litters contain few young, usually between two and five. There are exceptions: the Cape mole-rat produces up to 10 young and the Naked mole-rat has given birth to as many as 27 in captivity, though the average litter size is 12. It must be remembered, however, that the Cape mole-rat tends to surface more often than other species and is therefore more liable to be preyed upon, and so may produce more young to compensate for exceptional loss, and that in a Naked mole-rat colony only one female breeds.

All genera, except Naked mole-rats, appear to be seasonal breeders, having one or two litters during the breeding season. In captivity the breeding female Naked mole-

The young born to the colony are cared for by all the mole-rats but suckled only by the breeding female. During weaning (from three weeks old), in addition to eating food brought to the nest the young beg feces from colony members: among other things this appears to be important in providing nutrients for the young mole-rats. Once weaned they join the worker caste, but whereas some individuals appear to remain as workers throughout their lives, others eventually grow larger and become nonworkers. It is therefore likely that a colony is composed of a number of closely related litters—many with the same parents. As with social insects, this relatedness is probably an important factor in the evolution and maintenance of a social structure in which some individuals in the colony never breed. By caring for closely related mole-rats which share their genetic make-up, the nonbreeding individuals ensure the survival and passing on of their own genetic characteristics, even when they themselves do not actually breed.

This type of social system often evolves in animals in which an individual or pair has a poor chance of surviving on its own and of successfully rearing young. Under these circumstances it pays them to team up with other individuals and for some animals not to breed but to devote all their energies to ensuring that young born to their close relatives survive. This appears to be true for Naked mole-rats which live in arid regions and have to bore through very hard soil to find their food. By joining forces with a number of individuals and by channeling the

not sterile, why do they normally not breed? After all, individual mole-rats cannot appreciate the long-term evolutionary significance of not breeding, so what prevents them from breeding? It appears that the breeding female suppresses reproduction in her colony. As with the social insects, this control seems to be largely chemical (pheromones) produced by the breeding female. In captive colonies, with a history of many litters surviving to weaning, the whole colony is affected by the reproductive state of the breeding female. For example, just before a litter is born all the colony members develop teats and look like females: male hormone levels drop and some females come close to breeding condition. This strongly suggests that the colony is responding to chemical stimuli produced by the breeding female: in this case, the stimuli seem to prime the colony to receive and care for young which are not their own. The chemicals used and the ways in which the colony is exposed to them are currently under investigation.

Two sites of control are strongly implicated by experimental evidence. The first is through chemicals in the urine of the breeding female which is deposited in the communal toilet area and becomes spread over other colony members. The second site of possible control is in the communal nest where there is very close body contact. This would be an ideal situation for transmitting chemical signals to the colony.

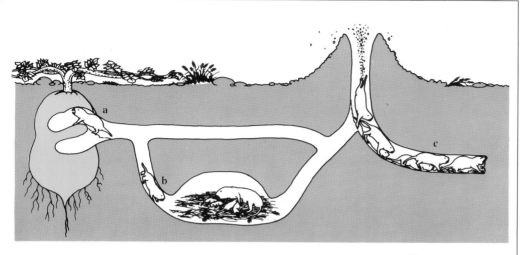

THE NAKED MOLE-RAT UNDERGROUND ORGANIZATION

▲ **Cross-section of a burrow system.** On the left (**a**) Naked mole-rats hollow out a growing tuber; in the center (**b**) is the main chamber which is occupied by the breeding female, subsidiary adults and young; on the right (**c**) a digging chain is at work.

◄ **A digging chain of Naked mole-rats.** The front mole-rat excavates with its teeth and pushes the soil backwards. The animal behind pulls it behind him and then moves backwards, holding himself close to the tunnel floor, until it can pass the soil to the animal responsible for dispersal. It then returns to the front, straddling the line of mole-rats pushing soil away.

► **The appearance of Naked mole rats** varies according to function. From top to bottom: (**i**) breeding male; (**ii**) breeding female; (**iii**) nonworking adult; (**iv**) working adult.

▼ **A young Naked mole-rat** in the huddle.

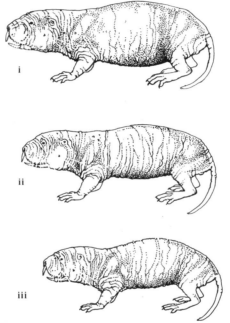

rat produces a litter every 80 days (just over 4 litters a year).

The only predators known to be able to pursue mole-rats underground are snakes. Field evidence suggests that the Mole snake (*Pseudaspis cana*) is attracted to the smell of freshly thrown up soil and will penetrate the burrow system via a new molehill. This may also be true of the Eastern beaked snake (*Rhamphiophis oxyrhunchus rostratus*) which has been seen preying on Naked mole-rats as they kick soil out of the burrow. Other predators take mole-rats when they venture above ground or are working very near the surface; mole-rat skulls are not uncommon in the pellets of birds of prey.

In addition to protecting the mole-rat against predators, the underground environment provides a uniform humid microclimate. This and the high mosture content of the mole-rats' food precludes the necessity of having to drink free water. The burrow temperature remains stable throughout the day, often in stark contrast to the surface temperature. In Naked mole-rat country, for example, surface temperatures of over 140°F (60°C) have been recorded while burrows (8in (20cm), below ground) remained a steady 84–86°F (29–30°C). In response to this stable burrow temperature Naked mole-rats have almost lost the ability to regulate their body temperature which consequently remains close to that of the burrow. If they need to alter their body temperature they huddle together when cold and take refuge in cooler areas within the system if they overheat when, for example, digging close to the surface. Their naked skin permits a rapid transfer of heat between them and the environment. Because Naked mole-rats have low body temperatures their metabolic rates are also low and such things as their need for food and their rate of growth are less than normal.

Though inconspicuous animals, molerats can cause considerable damage to human property. They have the ability to chew through underground cables and to undermine roadways, and sometimes devour root crops. Even their hills of excavated soil can damage the blades of harvesting machines, not to mention garden lawns and golf courses. Recent studies, however, have shown that mole-rats are important agents in soil drainage and soil turnover (a single Cape dune mole-rat may throw up as much as 1,100lb (500kg) of soil each month). They may play a role is dispersing geophytes (plants with underground storage organs) and may eat plants that are poisonous to farm livestock.   JUMJ

LAGOMORPHS

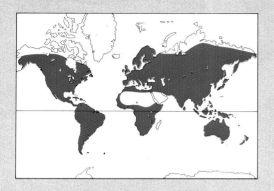

ORDER: LAGOMORPHA
Two families; 11 genera; 58 species.

Pikas
Family: Ochotonidae–ochotonids
Fourteen species in 1 genus (*Ochotona*).

Rabbits and hares
Family: Leporidae–leporids
Forty-four species in 10 genera.
Includes: **Antelope jack-rabbit** (*Lepus alleni*),
Black-tailed jack-rabbit (*L. californicus*),
European hare (*L. europaeus*), **Arctic hare**
(*L. timidus*), **Snowshoe hare** (*L. americanus*),
Sumatran hare (*Nesolagus netscheri*), **European
rabbit** (*Oryctolagus cuniculus*), **Volcano rabbit**
(*Romerolagus diazi*), **Swamp rabbit** (*Sylvilagus
aquaticus*), **Forest rabbit** (*S. brasiliensis*), **Marsh
rabbit** (*S. palustris*), **Eastern cottontail**
(*S. floridanus*).

WHAT do the mad "March hare" and the diminutive rock coney or pika have in common; and how do they relate to the pestilential European rabbit and the rare Volcano rabbit? They are all lagomorphs, which literally means "hare-shaped."

Lagomorphs occur throughout the world either as native species or introduced by man. The order contains two families: the small rodent like pikas (family Ochotonidae) which weigh less than 1.1lb (0.5kg); and the rabbits and hares (family Leporidae), the largest of which may weigh over 11lb (5kg). Pikas are small with short front and hind legs and are well adapted for living in rocky habitats where many species are found; the tail is virtually absent (at least externally) and their ears are short and rounded. Rabbits and hares, on the other hand, have a more flattened body, and elongate hind-limbs adapted for running at speed over open ground. Their ears are long and mobile and the tail is usually a small "powder puff."

Lagomorphs, whether hares or pikas, have characteristics which give them a similar appearance. Their fur is usually long and soft and their feet, unlike those of many rodents, are fully furred; their ears are large (even pikas have ears larger than most rodents) and their eyes are set high on their heads and look sideways giving them a wide field of vision. The nose has slit like nostrils which can be opened and closed by a fold of skin above—thus rabbits are often said to "wink" their noses. In stature lagomorphs have weak but flexible necks which enable them to turn their heads more than most rodents and when resting they usually tuck their heads back into their shoulders. Lagomorphs have only one external opening (cloaca) for both anus and urethra.

Lagomorphs are herbivores feeding on grasses, leaves and bark, as well as seeds and roots, even though some have been reported as eating snails and insects—even ants in the case of the Eastern cottontail. The lagomorph digestive system is highly modi-fied for coping with large quantities of vegetation; particularly they eat some of their feces (see diagram).

Lagomorphs were originally classified as rodents (order Rodentia) because of their gnawing incisors and herbivorous diet. In 1912 J. W. Gridley recognized several distinctive features which set them apart from rodents and he created the new order Lagomorpha. Unlike rodents, lagomorphs possess a second pair of small, peg-like incisors behind the long constantly growing pair in the upper jaw. This gives rabbits and hares a dental formula of I2/1, C0/0, P3/2, M3/3; pikas have one fewer upper molar on each side. Serological studies indicate that lagomorphs are no more closely related to rodents than to any other mammal order.

The oldest known lagomorphs occur in the late Eocene (50 million years ago) in Asia and North America, but the more primitive Asian forms may pre-date the first North American records. Despite probable Asian origins, the family Leporidae underwent most of its Oligocene and early Miocene development in North America (38–20 million years ago). The first pikas appeared in Asia in the middle Oligocene and spread to North America and Europe in the Pliocene (7–2 million years ago). Pikas seem to have peaked in distribution and diversity during the Miocene (26.7 million years ago) and since declined while the rabbits and hares have maintained a widespread distribution since the Pliocene. Evolutionary changes in both families have been conservative.

Rabbits and hares have nevertheless occupied a remarkably wide range of habitats; these include the cold snow-covered arctic habitat of the Arctic and Snowshoe hares, the semi-desert habitat of the Antelope and Black-tailed jack rabbits, the grassland habitat of the European hare, the tropical forest of the South American Forest rabbit, the tropical mountain forest of the Sumatran hare and the swamps of the

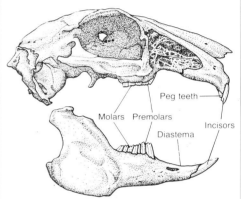

▲ **Rabbit skull.** Lagomorphs have long, constantly growing incisors, as do rodents, but lagomorphs differ in having two pairs of upper incisors, the back non-functional ones known as peg teeth. There is a gap (diastema) between the incisors and premolars. The dental formula of rabbits and hares (family Leporidae) is I2/1, C0/0, P3/2, M3/3, with the pikas (family Ochotonidae) having one less upper molar in each jaw.

▶ **Motionless in grass,** a young European hare (leveret) hides in a form awaiting its mother's return. European hare leverets are dispersed to such separate forms about three days after birth.

▶ **Double digestion.** The lagomorph digestive system is highly modified for coping with large quantities of vegetation. The gut has a large blind-ending sac (the cecum) between the large and small intestines which contains a bacterial flora to aid the digestion of cellulose. Many products of the digestion in the cecum can pass directly into the blood stream, but others such as the important vitamin B_{12} would be lost if lagomorphs did not eat some of the feces (refection) and so pass them through their gut twice. As a consequence lagomorphs have two kinds of feces. Firstly, soft black viscous cecal pellets which appear during the day in nocturnal species and during the night in species active in the daytime. These are usually eaten directly from the anus and stored in the stomach, to be mixed later with further food taken from the alimentary mass. Secondly round hard feces which are passed normally.

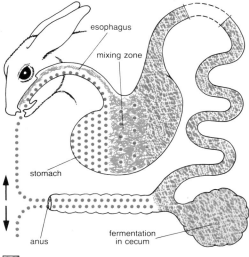

esophagus

mixing zone

stomach

anus

fermentation
in cecum

first passage of food   ••• cecal pellets

alimentary mass   ••• hard feces

North American Marsh and Swamp rabbits. Pikas, on the other hand, are highly adapted to their alpine habitat of North America and Asia, and steppes in Asia.

Lagomorphs represent staple prey to a range of medium-sized mammalian and bird, and larger predators. In species like the European and cottontail rabbits, predation, disease and climatic factors are the primary agents responsible for mortality rates often in excess of 90 percent among young of the year. High mortality rates are counteracted by a well-recognized high reproductive capacity. Most species reach sexual maturity relatively early (three months in female European rabbits). The gestation period is short: 40 days in *Lepus* species and about 30 days in all other members of the order; litter size is often large. Other features of lagomorph reproduction which minimize inter-birth interval for females include the phenomenon of induced "ovulation", where eggs are shed in response to copulation rather than only a cyclic basis; and "postpartum estrus" where a female is able to conceive immediately after giving birth. Female lagomorphs are also capable of "resorbing" embryos under adverse conditions, for example climatic or social stress; this clearly reduces the energetic loss of premature pregnancy termination for females. There is also evidence that some species, like the European hare, are capable of conceiving a second litter before birth of the last young—"superfetation"; explanations as to just how a female is able to support embryos at different stages of development present reproductive physiologists with something of a challenge!

JAC/ES

RABBITS AND HARES

Family: Leporidae
Forty-four species in 10 genera.
Distribution: Americas, Europe, Asia, Africa; introduced in Australia, New Zealand and other islands.

Habitat: wide ranging, from seashore to upper mountainous regions, from arctic tundra to city center, from dry desert to the swamp, from agricultural landscape to forest.

Size: ranges from head-body length 10–11.4in (25–29cm), weight 0.66lb (0.3kg) for Pygmy rabbit, and tail length 0.6in (1.5cm) (Amami rabbit) to head-body length 20–30in (50–76cm), weight 5.5–11lb (2.5–5kg) and tail length 2.8–4.7in (7–12cm) in European hare. Ear length ranges from about 1.7in (4.3cm) in Sumatran hare to 7–8in (18–20cm) in the Antelope jackrabbit; ears longer than wide, Hind limbs longer than forelimbs. Y-shaped naked groove extends from the upper lip to and around nose ("hare-lip").

Coat: usually thick and soft, but coarse in some forms; hair on the ears shorter and thinner; tail well furred or even bushy; feet hairy on both surfaces; coloration ranges through reddish brown, brown, buff, gray, or white, belly often covered with lighter or pure white hair. One species striped, and arctic forms change into white for winter.

Gestation: usually longer in hares (Mountain hare 50 days) than in rabbits (European rabbit 28–33 days).

Longevity: average less than 1 year in the wild; physiological maximum age of European hare is estimated 12–13 years; oldest recorded age in the wild is 12.5 years.

Species include: **Bushman hare** (*Bunolagus monticularis*); **Hispid hare** (*Caprolagus hispidus*); **Antelope jackrabbit** (*Lepus alleni*); **Black-tailed jackrabbit** (*L. californicus*); **Snowshoe hare** (*L. americanus*); **European hare** (*L. europaeus*); **Arctic hare** (*L. timidus*); **Sumatran hare** (*Nesolagus netscheri*); **European rabbit** (*Oryctolagus cuniculus*); **Amami rabbit** (*Pentalagus furnessi*); **Volcano rabbit** (*Romerolagus diazi*); **Forest rabbit** (*Sylvilagus brasiliensis*); **Eastern cottontail** (*S. floridanus*).

THERE are few mammals so involved with man as the European rabbit. Domestication of the species started probably in Roman times in northern Africa or Italy; today there are over 50 established strains of domestic rabbit all selectively bred from this one species. There have been many deliberate worldwide introductions of the European rabbit to countries like Australia by 19th-century empire-builders (seeking to establish the fauna and hunting from home), and to scattered islands throughout the oceans as a food source for shipwrecked sailors. Most of the successful mainland invasions have been by wild-type stock; colonies of domesticated rabbits only managed to survive on islands without heavy predation pressures. In the absence of those natural selection pressures (predators, climate and disease) operating in their ancestral home, many of these introduced populations exploded to reach the pest proportions familiar in Australia in the first half of this century.

Yet the European rabbit is just one species of a family (Leporidae) of 44 species, some of which have similar but less notorious involvement with man, and some whose numbers can be counted possibly in hundreds rather than tens of millions.

The family Leporidae can be broadly divided into two groups: the jackrabbits and hares (genus *Lepus*); and the rabbits in the remaining 9 genera. Although many of the latter genera include species popularly known as "hares", for example the Hispid hare, they look more like rabbits.

All rabbits and hares are adapted for quick movement and flight from danger. Their hind legs are long and adapted for running—in full flight some of the larger hares can reach speeds of 50mph (80km/h). Their ears are large and mobile, to facilitate the detection of approaching danger. Unlike those of pikas (the other family of the order Lagomorpha), the eyes of rabbits and hares are large and adapted to their twilight and night-time activities.

The forefeet have five toes and the hind-feet four, all equipped with strong claws. The lower surface of the feet is covered with long, thick brush-like hairs which provide a better grip and cushion when moving on hard ground. This is especially true in the Snowshoe hare where it also helps the animal to run on top of snow.

Many of the differences between hares and rabbits are associated with specializations for a burrowing life (most rabbits) or improved running ability (hares). A few hare species are known to burrow in extreme climates, eg the Black-tailed jackrabbit digs short burrows to avoid high summer temperatures in the desert. Snowshoe and Arctic hares may burrow into snow. In Scotland, Arctic hare populations often dig burrows 3.3–6.6ft (1–2m) long which are used as bolt holes by young—but rarely used by adults. The same species is known to dig complex 23ft (7m) tunnels in Khatanga (Russia), but doesn't appear to burrow in other countries, for example the Swiss Alps or Ireland. Forms—surface depressions in the ground or vegetation—are more commonly used as resting up sites by hares. These may be well-established sites used by successive generations or temporary refuges occupied for a few hours. The underground network of tunnels and chambers dug by rabbits show considerable variety in form, depending on species and within species according to habitat (see pp140–141).

The occurrence of rabbits and hares in different climatic conditions is reflected in

8

Representative species of rabbits and hares.
(1) The Greater red rockhare (*Pronolagus crassicaudatus*) in an alert scanning posture.
(2) The Hispid hare (*Caprolagus hispidus*), sitting among cuttings and pellets. (3) The European hare (*Lepus europaeus*), boxing. (4) The Volcano rabbit (*Romerolagus diazi*) reingesting pellets, sitting amongst vegetation of *zacatón* grasses.
(5) A male Eastern cottontail (*Sylvilagus floridanus*) in an alert posture. (6) The Sumatran hare (*Nesolagus netscheri*), grooming its muzzle and spreading scent. (7) The European rabbit (*Oryctolagus cuniculus*), a dominant male chin rubbing. (8) An Amami rabbit (*Pentalagus furnessi*) digging its burrow. (9) A Bushman hare (*Bunolagus monticularis*) in an alert posture. (10) A Bunyoro rabbit (*Poelagus marjorita*) hopping.

their molting physiology. Different species undergo one to three regular seasonal changes of fur, in some cases resulting in sharp differences between summer and winter coloration. Molting is triggered by temperature and light intensity. The Arctic hare never attains a white winter coat in Ireland but retains it for five months of the year in the European part of the USSR and for seven months in some northern Asian regions.

Young hares (leverets) are born, at a form site, at a more advanced stage of development than rabbit young (kittens). The latter are generally born at carefully constructed nests (of hair and grass) within the warren or special shallower breeding "stop" sites. Leverets are covered with fur at birth and their eyes are open; rabbit kittens are naked and their eyes do not open for several days (10 in the European rabbit). Those species studied so far appear to suckle their litters for only one brief period, typically under 5 minutes, once every 24 hours. (The milk of female European rabbits, for example, is known to be highly nutritious and with 10 percent fat and 15 percent protein it is richer than goat or cow milk.) For female European rabbits this will be the only contact between

mother and young until weaning (around 3 weeks of age); those litters born at breeding stop nests, for instance, will be re-enclosed into their soil-covered tunnel between each nursing. European hare leverets are dispersed to separate forms about 3 days after birth, but meet up daily at a specific location at a set hour (around sunset) for less than 3 minutes nursing. Radio-tracked litters of Snowshoe hare showed a similar behavior.

Puberty is attained in approximately 3–5 months in the European rabbit and about 1 year in most hare species; however some species of hares take two years.

The relationship between climate and reproduction is illustrated by the New World rabbits in the genus *Sylvilagus*. They show a direct correlation between latitude and litter size; species or subspecies in the north produce the largest litters which are generally correlated with the shortest breeding season. There is also a relationship between latitude and the length of the gestation period—rabbits in northern latitudes have the shortest gestation periods. The advantage of a short gestation is that the maximum number of young can be produced during the period when the weather is most

▲ **Big ears**—a Black-tailed jackrabbit in its hot, arid habitat. It is active at night, the day being spent in the shade. Its ability to regulate the flow of blood through the massive ears controls the intake or loss of heat according to environmental conditions.

▶ **The most abundant** and best-studied of cottontails, ABOVE, the Eastern cottontail, is found in almost all types of habitat and mature woodland. Young (kittens) are born on the surface, are blind and have only a sparse covering of hair.

▶ **Seeking shade under a cactus,** a Desert cottontail rests during the daytime heat. Most cottontails are active at night or twilight, but may be seen during the day at any time.

▷ **Jackrabbit in snow.** OVERLEAF As is common to several species from northern regions, the coat molts in the fall and is replaced by a thicker, paler (often white) one which aids heat conservation and camouflage in the snowy wastes.

suitable. Conversely, it is advantageous for rabbits to have longer gestation periods in southern localities because young rabbits born more fully developed are better able to avoid predators and fend for themselves.

The potentially high reproductive capacity of rabbits and hares means populations can have a rapid rate of increase. Population cycles have been observed in some species, particularly in the Snowshoe hare (see pp138–139).

Rabbit social behavior varies from the highly territorial breeding groups of the European rabbit to (see pp140–141) the non-territorial Eastern cottontail.

The only auditory signals known for most species are the characteristic foot thumps—made in alarm or aggression—and the distress screams made by captured young and adults. Some species like the Volcano rabbit are reported to use a variety of calls. Scent signals seem to play a predominant role in the communication systems of most rabbits and hares.

Because of their abundance, rabbits and hares are widely interwoven into man's economy. They appear frequently in the art of ancient civilizations, for example of the Central American Mayans, and were regarded as prime sporting animals for the "hunt" by Europeans. To the Romans, roast rabbit meat and embryos—laurices—were a delicacy, with the result that they kept rabbits and hares in walled enclosures—"leporaria." The good quality of its light, soft meat made the European rabbit a valued source of food. During the Middle Ages rabbits were kept in the enclosures belonging to monasteries and many of today's populations of wild European rabbits are descendants of escapees. Rabbit meat remains an important source of protein but is less popular in, for example, the United Kingdom since myxomatosis.

Modern agriculture involving deforestation and grazing by livestock, and the elimination of predators, created conditions (open grassland, cultivated crops and few predators) favorable for rabbits (the European rabbit is preyed upon by over 40 vertebrate species in its ancestral home in the Iberian peninsula, but much fewer in the rest of Europe). In early postglacial times the European hare inhabited the open steppes of eastern Europe and Asia, and it is likely that few occurred in the limited open habitats of central and western Europe. Advancing agriculture later replaced the open forests with a patch-work of cultivated fields, meadows, pastures and isolated woodlands and hedgerows, and a profusion of "new" wild and cultivated plants emerged—ideal hare habitat. The European hare rapidly moved into this habitat and is now found throughout Europe.

In Australia, introduced European rabbits caused enormous damage to both virgin habitats and cultivated areas between 1900–1959. The European hare is regarded as a pest in Australia and New Zealand where it has been introduced. This species was also introduced to Argentina (from Germany) in 1888 at Rosaria Santa Fe province. Since then it has spread throughout the country from Patagonia to the subtropical north covering an area of around 1 million sq mi (2.7 million sq km). Densest populations occur in the "humid pampas" area where 5–10 million hares are caught each year with nearly 15,000 tons of meat exported to Europe in 1977.

In the United States the Black-tailed jackrabbit of California is the most serious pest of cultivated crops, while the Snowshoe hare and cottontails cause greatest disruption of the reforestation program through their destruction of tree seedlings.

Large-scale rabbit drives were a popular method of control at the beginning of the 20th century, but modern control programs use a combination of techniques including

poison baits, buffer crops, repellents, exclusion fencing, shooting and trapping. Control of European rabbits in Australia was achieved by the introduction of the virus disease myxomatosis. This is harmless to its natural host the Forest rabbit or tapiti of South America and to other members of the genus *Sylvilagus*, but is virulent and deadly to the European rabbit. Following the introduction of myxomatosis into the wild rabbits of Australia in 1951–52, massive numbers died, relieving the country of a devastating pest. A similar decline occurred in the United Kingdom and other European countries. But now both in Europe and Australia rabbits are beginning to develop a level of immunity to the disease and their numbers have again increased.

In North America the transplantation of eastern forms to the west coast has caused changes in the genetic make-up of resident forms. For example, with a decreasing population of the native subspecies of the Eastern cottontail, wildlife agencies, hunt clubs and private individuals started, in the 1920s, a massive cottontail importation program into Kansas, Missouri, Texas and Pennsylvania. The scheme failed in that the annual harvest remained low, but one incidental consequence is that the native subspecies has been replaced by its hybrid with the imported one, which is colonizing new habitats.

While some rabbits and hares have reached pest status others are in danger of extinction. In general these are relict species, often the only members of their genus, specialized to a restricted threatened habitat. The Sumatran hare, of which only 20 animals have ever been recorded, comes from remote mountainous regions of Sumatra. The Hispid hare found in scattered pockets of sal forests in northern India and Bangladesh is losing the fight to compete with humans who are either burning its habitat to improve grazing or using its food plants as thatch. The Bushman hare inhabits riverside habitats that are rapidly being cultivated. The Volcano rabbit is restricted to the volcanic slopes near Mexico City, that is within a 30-minute drive of 17 million people. As well as habitat destruction this endangered species suffers the pressures of tourism and hunting. The Amami rabbit is found only on two heavily forested islands in the Amami Island group of Japan. Incessant logging has reduced the population to about 5,000 and its endangered status is supported by its creation in Japan as a "special natural monument."

JAC/ES

THE 44 SPECIES OF RABBITS AND HARES

Genus *Bunolagus*

Bushman hare [E]
Bunolagus monticularis
Bushman or River hare.

Central Cape Province (S Africa).
Dense riverine scrub (not the
mountainous situations often
attributed). Now extremely rare.
Coat: reddish, similar to Red
rockhares with bushy tail.

Genus *Caprolagus*

Hispid hare [E] [*]
Caprolagus hispidus
Hispid hare or Assam rabbit.

Uttar Pradesh to Assam; Tripura
(India), Mymensingh and Dacca on
the western bank of river
Brahmaputra (Bangladesh). Sub-
Himalayan sal forest where grasses
grow up to 8ft in height during the
monsoon months; occasionally
cultivated areas. HBL 19in; TL
2in; EL 3in; HFL 4in; wt
5.5lb. Coat: coarse and bristly;
upperside appears brown from
intermingling of black and brownish
white hair; underside brownish white
with chest slightly darker; tail brown
throughout, paler below. Claws
straight and strong. Inhabits burrows
which are not of its own making.
Seldom leaves forest shelter.

Genus *Lepus*

Hares
Most inhabit open grassy areas, but:
Snowshoe hare occurs in boreal
forests; European hare occasionally
forests; Arctic hare prefers forested
areas to open country; Cape hare
prefers open areas, occasionally
evergreen forests. Rely on well-
developed running ability to escape
from danger instead of seeking cover;
also on camouflage by flattening on
vegetation. HBL 16–30in; TL
1–5in; wt 3–11lb. Coat:
usually reddish brown, yellowish
brown or grayish brown above,
lighter or pure white below; ear tips
black edged with a significant black
area on the exterior in most species;
in some species the upperside of the
tail is black. Indian hare has a black
nape. Species inhabiting snowy
winter climate often molt into a white
winter coat, while others change from
a brownish summer coat into a
grayish winter coat. Diet: usually
grasses and herbs, but cultivated
plants, twigs, bark of woody plants

are the staple food if others are not
available. Usually solitary, but
European hare more social. Habitat
type has a marked effect on home-
range size within each species, but
differences also occur between
species, eg from 10–50 acres in Arctic
hares to over 750 acres in European
hares. Individuals may defend the
area within 3–6ft of forms but home
ranges generally overlap and feeding
areas are often communal. Most live
on the surface, but some species, eg
Snowshoe and Arctic hares dig
burrows while others may hide in
holes or tunnels not of their making.
Breed throughout the year in
southern species; northern species
produce 2–4 litters during spring and
summer. Litter size from 1–9.
Gestation up to 50 days in Arctic
hare, other species shorter.
Vocalization: deep grumbling; shrill
calls given in pain. Twenty-one
species.

Antelope jackrabbit
Lepus alleni

S New Mexico, S Arizona to N Nayarit
(Mexico), Tiburon Is. Locally
common. Avoid dehydration in hot
desert by feeding on cactus and
yucca.

Black-tailed jackrabbit
Lepus californicus

Mexico, Oregon, Washington,
S Idaho, E Colorado, S Dakota,
W Missouri, NW Arkansas, Arizona,
N Mexico. Locally common.

White-sided jackrabbit
Lepus callotis

SE Arizona, SW New Mexico and
Oaxaco (Mexico). Locally common,
but declining.

Tehuantepec jackrabbit
Lepus flavigularis

Restricted to sand dune forest on
shores of salt water lagoons on
nothern rim of Gulf of Tehuantepec,
(Mexico). Nocturnal.

Black jackrabbit
Lepus insularis

Espiritu Santo Is (Mexico).

White-tailed jackrabbit
Lepus townsendii

S British Columbia, S Alberta,
SW Ontario, SW Wisconsin, Kansas,
N New Mexico, Nevada, E California
Locally common.

Snowshoe hare
Lepus americanus

Alaska, coast of Hudson Bay,
Newfoundland, S Appalachians,
S Michigan, N Dakota, N New Mexico,
Utah, E California. Locally common.

Japanese hare
Lepus brachyurus

Honshu, Shikoku, Kyushu, (Japan).
Locally common.

Cape hare
Lepus capensis

Africa, S Spain (?), Mongolia,
W China, Tibet, Iran, Arabia. Locally
common.

European hare
Lepus europaeus
European or Brown hare.

S Sweden, S Finland, Great Britain
(introduced in Ireland), Europe south
to N Iraq and Iran, W Siberia. Locally
common but declining.

Savanna hare
Lepus crawshayi

S Africa, Kenya, S Sudan; relict
populations in NE Sahara. Locally
common.

Manchurian hare
Lepus mandshuricus

Manchuria, N Korea, E Siberia. Range
decreasing.

Ethiopian hare
Lepus starcki

Ethiopia.

Indian hare
Lepus nigricollis
Indian or Black-naped hare.

Pakistan, India, Sri Lanka (introduced
into Java and Mauritius.

Woolly hare
Lepus oiostolus

Tibetan plateau and adjacent areas.

Burmese hare
Lepus peguensis

Burma to Indochina and Hainan
(China).

Scrub hare
Lepus saxatilis

S Africa, Namibia.

Chinese hare
Lepus sinensis

SE China, Taiwan, S Korea.

Arctic hare
Lepus timidus
Arctic, Mountain or Blue hare.

Alaska, Labrador, Greenland,
Scandinavia, N USSR to Siberia,
Hokkaido, Sikhoto Alin Mts, Altai,
N Tien Shan, N Ukraine, Lithuania.
Locally common. Isolated populations
in the Alps and Ireland.

African savanna hare
Lepus whytei

Malawi. Locally common.

Yarkand hare
Lepus yarkandensis

SW Sinkiang (China). Rare.

Genus *Nesolagus*

Sumatran hare [*]
Nesolagus netscheri
Sumatran hare or Sumatran short-eared rabbit.

W Sumatra (1°–4°S) between
1,900–4,600ft in Barisan range.
Primary mountain forest.
HBL 14–15in; TL 0.6in;
EL 1.5–2in. Coat: variable; body
from buffy to gray, the rump bright
rusty with broad dark stripes from the
muzzle to the tail, from the ear to the
chin, curving from the shoulder to the
rump, across the upper part of the
hind legs, and around the base of the
hind foot. Diet: juicy stalks and leaves.
Strictly nocturnal; spends the day in
burrows or in holes (not of its own
making). Very rare—only one
specimen recorded in last decade.

Genus *Oryctolagus*

European rabbit
Oryctolagus cuniculus

Endemic on the Iberian peninsula and
NW Africa; introduced in rest of
W Europe 2,000 years ago, and to
Australia, New Zealand, S America
and some islands. Opportunistic,
having colonized habitats from stony
deserts to subalpine valleys; also
found in fields, parks and gardens,
rarely reaching altitudes of over
2,000ft. Very common. HBL
15–20in; TL 1.5–3in; EL
2.5–3in; HFL 3–4.5in; wt
3–6.5lb. Coat: grayish with a fine
mixture of black and light brown tips
of the hair above; nape reddish-
yellowish brown; tail white below;
underside light gray; inner surface of
the legs buffy gray; total black is not
rare. All strains of domesticated rabbit
derived from this species. Colonial

organization associated with warren systems. Diet: grass and herbs, roots and the bark of trees and shrubs, cultivated plants. Breeds from February to August/September in N Europe; 3–5 litters with 5–6 young, occasionally up to 12; gestation 28–33 days; young naked at birth; weight about 1–1.5oz, eyes open when about 10 days old. Longevity: about 10 years in wild. Vocalization: shrill calls are given in pain or fear.

Genus *Pentalagus*

Amami rabbit E
Pentalagus furnessi
Amami or Ryukyu rabbit (erroneously).

Two of the Amami Islands (Japan). Dense forests. HBL 17–20in; EL 1.7in. Coat: thick and woolly, dark brown above, more reddish below. Claws are unusually long for rabbits at 0.3–0.7in. Eyes small. Nocturnal. Digs burrows. 1–3 young are born naked in a short tunnel; two breeding seasons.

Genus *Poelagus*

Bunyoro rabbit
Poelagus marjorita
Bunyoro rabbit or Uganda grass hare.

S Sudan, NW Uganda, NE Zaire, Central African Republic, Angola. Savanna and forest. Locally common. HBL 17–20in; TL 2–2.5in; wt 4–7lb; EL 2–2.5in. Coat: stiffer than that of any other African leporid; grizzled brown and yellowish above, becoming more yellow on the sides and white on the under parts; nape reddish yellow; tail brownish yellow above and white below. Ears small; hind legs short. Nocturnal. While resting, hides in vegetation. Young reared in burrows and less precocious than those of true hares. Said to grind teeth when disturbed.

Genus *Pronolagus*
Red rockhares
HBL 14–20in; TL 2–4in; HFL 3–4in; EL 2.5–4in; wt 4.5–6lb. Coat: thick and woolly, including that on the feet, reddish. Inhabits rocky grassland, shelters in crevices. Nocturnal, feeding on grass and herbs. Utters shrill vocal calls even when not in pain. Three species.

Greater red rockhare
Pronolagus crassicaudatus
S Africa.

Jameson's red rockhare
Pronolagus randensis
S Africa, E Botswana, Zimbabwe, Namibia.

Smith's red rockhare
Pronolagus rupestris
South Africa to Kenya.

Genus *Romerolagus*

Volcano rabbit E *
Romerolagus diazi
Volcano rabbit, teporingo or zacatuche.

Restricted to two volcanic sierras (Ajusco and Iztaccihuatl-Popocatepetl ranges) close to Mexico City. Habitat unique "zacaton" (principally *Epicampes*, *Festuca* and *Muhlenbergia*) grass layer of open pine forest at 9,000–13,000ft. Smallest leporid. HBL 11–14in; wt 16–20in; EL 1.5–2in. Features include short ears, legs and feet, articulation between collar and breast bones and no visible tail. Coat: dark brown above, dark brownish gray below. Lives in warren-based groups of 2–5 animals. Breeding season December to July; gestation 39–40 days; average litter 2. Mainly active in daytime, sometimes at night. Vocal like pikas.

Genus *Sylvilagus*
Cottontails
HBL 10–18in; TL 1–2in; wt 1–5lb; smallest Pygmy rabbit, biggest Swamp rabbit. Most species common. Coat: mostly speckled grayish brown to reddish brown above; undersides white or buffy white; tail brown above and white below ("cottontail"); Forest rabbit and Marsh rabbit have dark tails. Molts once a year, except Forest and Marsh rabbits. Ears medium sized (about 2in) and same color as the upper side; nape often reddish, but may be black. Range extends from S Canada to Argentina and Paraguay and a great diversity of habitats is occupied. Distributions of some species overlap. Most preferred habitat open or brushy land or scrubby clearings in forest areas, but also cultivated areas or even parks. Various species frequent forests, marshes, swamps, sand beaches, or deserts. Diet: mainly herbaceous plants, but in winter also bark and twigs. Only Pygmy rabbit digs burrows; others occupy burrows made by other animals or inhabit available shelter or hide in vegetation. Not colonial, but some species form social hierarchies in breeding groups. Active in daytime or night. Not territorial; overlapping stable home ranges of a few hectares. Vocalization rare. Longevity: 10 years (in captivity). Thirteen species. Most locally common.

Swamp rabbit
Sylvilagus aquaticus
E Texas, E Oklahoma, Alabama, NW–S Carolina, S Illinois. A strong swimmer. Gestation 39–40 days; eyes open at 2–3 days.

Desert cottontail
Sylvilagus audubonii
Desert or Audubon's cottontail.

C Montana, SW–N Dakota, NC Utah, C Nevada and N and C California (USA), and Baja California and C Sinaloa, NE Puebla, W Veracruz, (Mexico).

Brush rabbit
Sylvilagus bachmani
W Oregon to Baja California, Cascade–Sierra Nevada Ranges. Average 5 litters per year; gestation 24–30 days; covered in hair at birth.

Forest rabbit
Sylvilagus brasiliensis
Forest rabbit or tapiti.

S Tamaulipas (Mexico) to Peru, Bolivia, N Argentina, S Brazil, Venezuela. Average litter size 2; gestation about 42 days.

Mexican cottontail
Sylvilagus cunicularius
S Sinaloa to E Oaxaca and Veracruz (Mexico).

Eastern cottontail
Sylvilagus floridanus
Venezuela through disjunct parts of C America to NW Arizona, S Saskatchewan, SC Quebec, Michigan, Massachusetts, Florida. Very common. Gestation 26–28 days; young naked at birth.

Tres Marias cottontail
Sylvilagus graysoni
Maria Madre Is, Maria Magdalena Is, (Tres Marías Is, Navarit, Mexico).

Pygmy rabbit
Sylvilagus idahoensis
SW Oregon to EC California, SW Utah, N to SE Montana; isolated populations in WC Washington.

Omilteme cottontail
Sylvilagus insonus
Sierra Madre del Sur, C Guerrero (Mexico).

Brush rabbit
Sylvilagus mansuetus
Known only from San Jose Island, Gulf of California. Often regarded as subspecies of *S. bachmani*.

Mountain cottontail
Sylvilagus nuttalli
Mountain or Nuttall's cottontail.

Intermountain area of N America from S British Columbia to S Saskatchewan, S to E California, NW Nevada, C Arizona, NW New Mexico.

Marsh rabbit
Sylvilagus palustris
Florida to S Virginia on the coastal plain. Strong swimmer.

New England cottontail
Sylvilagus transitionalis
S Maine to N Alabama. Distinguished from overlapping Eastern cottontail by presence of gray mottled cheeks, black spot between eyes and absence of black saddle and white forehead.

ES

The Ten-year Cycle

Population fluctuations in the Snowshoe hare

Animal populations rarely, if ever, remain constant from year to year. Populations of the great majority of species fluctuate irregularly or unpredictably. An exception is the Snowshoe hare, whose populations in the boreal forest of North America undergo remarkably regular fluctuations which peak every 8–11 years. This is a persistent fluctuation, documented in fur-trade records for over two centuries and now widely known as "the 10-year cycle."

In addition to its regularity, the cycle is unusual in two other respects; firstly it is broadly synchronized over a vast mid-continental area from Alaska to Newfoundland, where regional Snowshoe hare peaks seldom differ by more than three years; secondly, the amplitude of change in numbers from a cyclic high to a low may be more than 100-fold.

There has been much speculation about what causes the Snowshoe hare cycle, but only recently have long-term field studies begun to provide the solution. It is now known that certain demographic events are consistently associated with each phase of the cycle. Thus declines from peak densities are initiated by markedly lower survival of young hares overwinter, and by sharp decreases in birth rates. These conditions persist for three or four years, and the population continues to contract. The survival of adult hares also declines during this phase of the cycle. The onset of the next cyclic increase in numbers is brought about by greatly improved rates of survival and birth. We know too that growth rates of young hares are highest during the increase phase of the cycle, and that overwinter weight losses are significantly lower at that time.

If Snowshoe populations fluctuate cyclically because of a pattern of changing survival and birth rates the next obvious question is what causes the latter? Acceptable explanations must account for the above-noted trends in juvenile growth rates and for weight losses overwinter, and for the cycle's synchrony between regions. There is currently no general consensus among biologists as to the ultimate cause of the 10-year cycle, but one tenable explanation runs as follows.

The cycle is repeatedly generated intrinsically when peak Snowshoe hare populations exceed their winter food supply of woody browse and resulting malnutrition triggers a population decline. As hare numbers fall, the ratio of predators to hares increases, as does the impact of predation on the hare population. This extends the cyclic

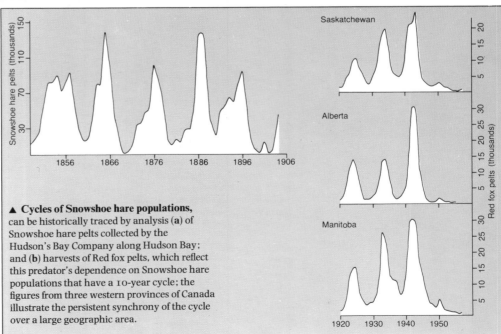

▲ **Cycles of Snowshoe hare populations,** can be historically traced by analysis (**a**) of Snowshoe hare pelts collected by the Hudson's Bay Company along Hudson Bay; and (**b**) harvests of Red fox pelts, which reflect this predator's dependence on Snowshoe hare populations that have a 10-year cycle; the figures from three western provinces of Canada illustrate the persistent synchrony of the cycle over a large geographic area.

► **Girdled by marauding hares.** Aspen trees in Alaska from which the bark has been chewed away by Snowshoe hares during winter. The thick snow cover enables the hares to reach far up the trunk. Snowshoe hares require such dense, low-growing woody vegetation for cover and winter food, while in the summer they eat a great variety of herbaceous plants.

◄ ▼ **A coat for each season.** Snowshoe hares have a gray-brown coat in summer BELOW, which turns pure white in winter LEFT. Some remain gray-brown year round in the humid coastal zones of western North America where snowfall is infrequent.

decline beyond the period of winter food shortage, and drives the hare population still lower. The resulting scarcity of hares then causes predator populations to drop to low levels. Now largely free of predation, and with winter food once more abundant, the Snowshoe population begins another cyclic increase. Inter-regional synchrony is caused by mild winters that moderate mortality and thereby delay declines of malnourished hare populations, while permitting others that are lagging to reach critical peak densities. Such synchrony is reinforced by highly mobile predators responding to local differences in hare abundance.

Among the more notable ecological consequences of the Snowshoe hare cycle are: firstly, the impact of hare browsing on plant species composition and succession—such unpalatable shrubs as the honey suckles (Caprifoliaceae) increase whereas forest succession from aspen (*Populus tremuloides*) to White spruce (*Picea glauca*) is slowed; secondly a direct and overriding effect on predator populations; thirdly, an indirect effect on the alternate prey of hare predators. The Snowshoe hare's significant influence on other components of the boreal forest ecosystem stems from the amplitude of its fluctuations, and the densities of 1,000 to 6,000 per sq mi (400–2,400 per sq km) commonly attained during cyclic peaks. While plant ecologists and physiologists have only recently acknowledged the impact of hare cycles on vegetation, the destabilizing effect on numbers of lynx, Red fox, coyote, North American marten, fisher and other predatory mammals has long been recognized by the fur trade. Populations of birds of prey likewise respond to fluctuating hare abundance, and together with the mammalian predators may thereby influence such alternate prey as Ruffed, Spruce and Sharp-tailed grouse.

Not all Snowshoe hare populations have a 10-year cycle. Those that do not are found where suitable habitat (a dense understory of woody shrubs and saplings) is highly fragmented or island-like. This fragmentation exists naturally in the mountain ranges and along the southern limit of Snowshoe distribution. In these regions predators have a greater diversity of prey, and hence more stable populations. Accordingly, the absence of cyclic fluctuations among Snowshoe hares has been ascribed to their being held in check by sustained predation, especially on individuals dispersing from habitat fragments. There is evidently a parallel between the Snowshoe hare in North America and the Arctic hare in Soviet Eurasia, for the latter also has a 10-year cycle within continuous habitat of the taiga, but exhibits irregular short-term fluctuations to the south as habitats become increasingly disjunct.

Population explosions are also seen in many rodents of the arctic tundra and taiga where they occur with regularity every 3–4 years, as in the lemmings. LBK

Habitat and Behavior

Adaptable females and opportunist males in rabbit societies

In Europe throughout the Middle Ages rabbits were farmed successfully for their meat and fur. They appeared to be tolerant of crowding and bred profusely in their small enclosures (called warrens) and lived together in underground burrows. Naturally enough they came to be regarded as a classic example of a mammal that lives in groups. The first systematic research into the nature of their social habits was conducted on similar enclosed populations in both Australia and England during the 1950s. The resulting observations were interpreted as showing that rabbits form mixed-sex breeding groups containing 6–10 adults, and defend exclusive group territories. Since then, however, some studies of wild populations have cast doubt on the general validity of these conclusions.

Two long-term studies in England have now shown that the European rabbit's habitat can have a profound effect on its social organization and behavior. One study took place on chalk downland, the other on coastal sand-dunes. At the former, the burrows in which rabbits typically take refuge for more than half of each day are clustered together in tight groups (also, confusingly, called warrens), which are themselves randomly distributed over the down. Adult females (does), who do most burrow excavation, rarely attempted more than the expansion of an existing burrow system in the hard chalky soil; completely new warrens hardly ever appeared. At the sand-dune site, by contrast, new burrows were continually being dug as others were collapsing or falling into disuse. Burrows were never found in the flat "slacks," which lie between the dunes and tend to flood easily, but even on the higher ground they were not clustered into the easily defined warrens found on the downland. Burrow entrances more than 16ft (5m) apart on the dunes were rarely connected by a tunnel and many had just a single entrance at the surface.

Does usually give birth and nurse their young in underground nesting chambers situated within pre-existing burrow systems. Thus, where space underground is in short supply natural selection should favor females that compete for and then defend some burrows. Not surprisingly there is some good evidence for such competition at the downland site. Of the disputes between adult females observed there, over 70 percent took place within 16ft (5m) of a burrow entrance habitually used by one of the contestants. There was also a direct relationship between the size of a warren and the number of adult females refuging in it: the larger the warren the more females lived there. Thus a group of females sharing a warren and feeding in extensively overlapping ranges around it are best regarded as reluctant partners in an uneasy alliance.

▲ ▶ Problems of burrowing. ABOVE European rabbits inhabiting sites with soft soil, for example sand dunes, have little difficulty digging new burrows, even overnight. BELOW RIGHT On hard soils, for example chalk, excavations are major endeavors so new burrows rarely appear. These differences considerably affect the social systems of rabbits on the two habitats; on sand dunes rabbits spread themselves out while those on chalk-land center on long-established burrow systems.

◀ Group-living and social behavior in European rabbits. Comparison of (**a**) dune-land and (**b**) chalk-land social organization.

On chalk-land rabbits have clustered burrows with females living as reluctant partners around each cluster; fights often occur between females and their home ranges overlap considerably within each group, but not with those of adjacent groups.

On dune-land rabbit burrows are not clustered and are randomly distributed, although they do not occur in the slacks which are prone to flooding. Females move freely between burrows and there is little fighting between individuals, and home ranges overlap less than on chalk-land.

In both habitats males have larger territories which overlap those of several females.

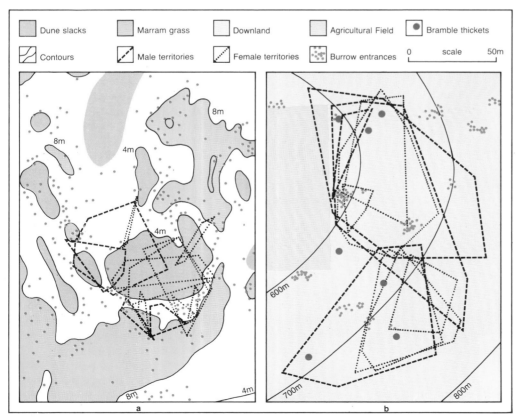

| Dune slacks | Marram grass | Downland | Agricultural Field | Bramble thickets |
| Contours | Male territories | Female territories | Burrow entrances | 0 scale 50m |

cent of interactions with does and refuged in the available burrows without hindrance from them. Males' home ranges were on average about twice the size of those of neighboring females with which they had an extensive overlap. Consequently these bucks could have been acquiring information, largely on the basis of scent rather than direct encounters, about the reproductive state of numerous females. Bucks did not seem to defend strict territories exclusive of all other males at either site. However, the frequent aggressive interactions observed between males, whether or not they were escorting does, may be best interpreted as attempts by them to curtail each other's use of space and access to females.

The behavior of bucks following females around can be regarded as "mate-guarding." Each female is usually accompanied by only one male, although males at the duneland site have been seen with up to three different females at different times over the course of a few days. If a second male approached a male-female pair he was promptly rebuffed by a look or a few paces in his direction by the consorting male. Occasionally energetic chases occurred, in which male antagonists jumped clear of the ground and attempted to rip each other with their claws as they passed in mid-air. Guarding seems to be effective: successful takeovers of paired females by approaching males were very rarely seen.

Despite these efforts by particular bucks to monopolize proximity to particular females, mating in rabbits is promiscuous. A recent Australian study involved the genetic typing (using blood proteins) of all potential parents in a population together with their weaned young. The resulting analysis showed that at least 16 percent of the young were not fathered by the male known from direct observation to be the usual escort of their mother. So, while promiscuous matings can yield offspring, the female's brief period in heat apparently makes it advantageous for bucks to try to monopolize sexual access to one female, or at most a few.

The dune- and downland studies in England have shown that, contrary to popular belief, rabbits are not always group-living animals. Sometimes females have to live together in order to make use of a limited supply of nesting sites, but this habit is not the automatic consequence of evolution. The evident flexibility of female social behavior contrasts with the similar behavior of the bucks in these two populations. As expected from the theory of sexual selection, male rabbits are sexual opportunists. PJG/DPC

At the duneland site burrows were never apparently in short supply; new ones often appeared overnight. Here, as expected, aggressive interactions between females were rare and not obviously related to burrow possession. Some does even moved round a series of "home burrows" during single breeding seasons without incurring attacks from local residents. This evidence of rather fluid refuging conventions, together with analyses showing that females' home ranges were not bunched together in superimposed clusters, suggests that adult does did not form groups in the dunes. Their burrowing activity was not constrained by the sandy substrate there, so they spread themselves out over the habitat.

What of the males? Buck rabbits, like many other male mammals, are unable to contribute directly to parental care. Consequently their reproductive success will reflect how many matings they have achieved with receptive females. Females come into a period of heat for 12–24 hours about every seventh day or soon after giving birth. Males apparently monitor female condition closely: adult does were "escorted" by single males for about a quarter of their time above ground in the breeding season on both study areas. Bucks won over 90 per-

PIKAS

Family: Ochotonidae

Fourteen species belonging to the genus *Ochotona*.

Distribution: N America, E Europe, Middle East, Asia N of Himalayas.

Habitat: rocky slopes on mountains, steppe and semidesert, ranging in height from sea level to 20,000ft (6,100m).

Size: ranges from head-body length 7.2in (18.3cm), weight 2.6–7.4oz (75–210g) in the Steppe pika to head-body length 8in (20.2cm), weight 6.7–10.2oz (190–290g) in the Afghan pika. Tail length 0.2–0.8in (0.5–2cm); ear length 0.5–1.4in (1.2–3.6cm).

Gestation: known for the Afghan pika, 25 days, and the North American pika, 30.5 days.

Longevity: 7 years for the North American pika, 4 years for the Asian pikas.

Species include: **Afghan pika** (*O. rufescens*); **Collared pika** (*O. collaris*); **Daurian pika** (*O. daurica*); **Large-eared pika** (*O. macrotis*); **Mongolian pika** (*O. pallasi*); **North American pika** (*O. princeps*); **Northern pika** (*O. hyperborea*); **Red pika** (*O. rutila*); **Royle's pika** (*O. roylei*); **Steppe pika** (*O. pusilla*).

Pikas are small lagomorphs with relatively large rounded ears and short limbs. They almost look like cavies (Guinea pigs). The tail is slight and barely visible. Pikas are little known because they live high up in the mountains or below ground in deserts.

The word pika and the generic name *Ochotona* are derived from vernacular terms used by Mongol peoples. There are probably 14 species of pikas, 12 distributed in Eurasia, two in North America. However, zoologists are unable to agree on an exact number of species.

Each species has a particular preference in habitat. Most prefer slopes covered in rock debris (talus) or rock slides on mountains. These rock dwellers move in crevices between loose rocks or among slide-rocks, lava flows or even the stone walls of houses. Of the rock dwellers the Afghan pika and the Mongolian pika also live in areas without rocks, in burrows. The Steppe pika and the Daurian pika inhabit steppes by constructing burrow systems, though their feet and nails are not specialized for burrowing. Species probably do not coexist in the same habitat. In Mongolia, the Daurian pika occupies steppes whereas the Mongolian pika inhabits rocky terrain. On high mountains, two rock dwellers segregate their habitats vertically. In the Nepal Himalayas the Large-eared pika lives at higher altitudes than Royle's pika, and in Tien-Shan higher than the Red pika. In Nepal the distribution boundaries of the two pikas border each other.

Pikas are usually active by day and do not hibernate. They are very lively and agile, but often sit hunched up on rocks for long periods. Most species are most active in the morning and late afternoon, with some activity at night. Remarkable differences are found in two Himalayan species: Royle's pika is active at dawn and dusk, but the Large-eared pika, living at altitudes over 13,000ft (4,000m) in Nepal, is active only in the hours around midday. The activity rhythms of these species are thought to result from living where there are favorable temperatures.

Pikas appear to utilize whatever plants are available near to their burrows or at the edges of their rock slides. They eat the leaves, stalks and flowers of grasses, sedges, shrub twigs, lichens and mosses. Pikas cannot grasp plants with their forepaws, so they eat grasses or twigs from the cut end. Pikas, like other lagomorphs, produce two different types of feces: small spherical pellets like pepper seeds, and a soft, dark-greenish excrement. The latter, having high energy value (particularly in B vitamins), is reingested either directly from the anus or after being dropped.

During summer and fall pikas spend much time collecting plants for winter food. They cut down green plants which rock dwellers carry to traditional places under rocks while burrow dwellers with dry habitats construct haypiles in the open near their burrows or between shrubs. Weights of haypiles vary but sometimes reach 13lb (6kg). Moreover, the Mongolian pika carries pebbles, 1–2in (3–5cm) in diameter, in its mouth and places them near its burrow, probably to prevent hay from being scattered by the wind. This hay-gathering

underground or running. In the mating season males of two North American species (the North American and Collared pikas) and of the Northern pika frequently give successive calls to declare their possession of territory. From summer to fall these pikas, both sexes, frequently give short calls, accompanied by the development of hay-gathering behavior. In contrast to these pikas two of the Himalayan species rarely utter even weak sounds and virtually abstain from hay-gathering. Parallel development between food storage and vocalization from summer to fall seems to indicate a possible causal relationship between these behavior patterns in evolution.

There are two different pika reproduction strategies, which are linked with habitat preferences. For typical rock dwellers the mean litter size is less than five and the young first breed as yearlings, but for burrow dwellers, and the Afghan pika and the Mongolian pika with their burrowing ability, the litter size is larger and the young born first breed in the summer of their first year. The relatively low reproductive capacity in the typical rock dwellers distributed at high altitudes or high latitudes may be related to the short favorable reproductive season. On the other hand, in the latter high reproductive capacity group, the densities greatly fluctuate (for example, from 0.8–1.2 to 28–32 animals per acre (2–3 to 70–80 per ha) for the Daurian pika) and the dispersed young may extend their habitats by burrowing.

Pikas always move alone. However, in the Northern pika a male and female hold the possession of a definite territory throughout the year. Each resident will react to intruders of the same sex, though females confine themselves to their territories and male residents often trespass into adjacent territories. The haypiles stored in each territory are consumed by the pair during the winter. In the mating season the two North American species possess a pair territory similar to that of the Northern pika, but in the hay-gathering season their pair territories are divided into two solitary territories, each on average covering 7,630sq ft (709sq m). Each protects the hay within its own "hay territory" from intruders of both sexes (see pp144–145). The social patterns of some Himalayan species from fall to winter are basically similar to the pair territoriality of the Northern pika. Thus the phenomenon of individual "hay territoriality" seems to be superimposed on the social organization of both North American species to ensure a supply of winter food. TKa

▲ ◄ **More like rats than rabbits** (their closest relatives), pikas have rounded ears, short limbs and no visible tail. The Collared pikas shown here (ABOVE, ABOVE LEFT) occur on rocky outcrops in Alaska and northwestern Canada. Their name is derived from the grayish patches below the cheek and around the neck. They spend about half of their time on the surface sitting on prominent rocks.

◄ **The Large-eared pika of Nepal** and surrounding areas occurs at heights between 7,500–20,000ft (2,300–6,100m). It has the largest ears of all pikas.

behavior is common to all species except the two Himalayan ones (Royle's pika and the Large-eared pika). In winter, in addition to eating their hay, pikas make tunnels in the snow to reach and gnaw the bark of apple trees, aspens and young conifers. Mongolian herders and antelopes steal the hay of Daurian pikas in winter. To compensate, this pika and Royle's pika steal wheat and fried wheat cakes from houses.

Most pikas give distinctive calls while perching on rocks, and occasionally while

Family Bonds and Friendly Neighbors
The social organization of the North American pika

Two North American pikas darted into and out of sight on a rock-strewn slope (talus), one in pursuit of the other. The chase continued, onto an adjoining meadow then into the dense cover of a nearby spruce forest. When next seen, dashing back towards the talus, they were being chased by a weasel. One pika was caught and death came quickly less than 3ft (about 1m) from the talus and safety. Immediately all pikas in the vicinity, with one exception, broke into a chorus of consecutive short calls, the sounds that pikas utter when alarmed by predators. The dead pika had initiated the chase, but the object of his aggression had escaped the weasel. He now surveyed the talus from a prominent rock perched in silence.

Most accounts of the natural history of pikas have emphasized their individual territoriality, but recent work in the Rocky Mountains of Colorado has revealed in detail their social organization. For example, adjacent territories are normally occupied by pikas of the opposite sex. Male and female neighbors overlap each other's home ranges more and have centers of activity that are closer to each other than are the ranges or activity centers of nearest neighbors of like sex. The possession and juxtapositions of territories tend to be stable from year to year. Pikas in North America can live up to 6 years, and the appearance and whereabouts of vacancies on the talus are unpredictable. For a pika, trying to secure a vacancy is like entering a lottery where in part an animal's sex determines whether it will have

▲ ► Collecting in the hay. A characteristic (almost frantic) activity of pikas in late summer is the harvesting of vegetation (RIGHT) to store in haypiles on the talus, in part to serve as food over winter. Most stores of hay are located under overhanging rocks (ABOVE).

◄ Singing pika. Pikas have two characteristic vocalizations: the short call and the long call (or song). Long calls (a series of squeaks lasting up to 30 seconds) are given by males primarily during the breeding season. Short calls normally contain one to two note squeaks and may be given: from rocky promontories either before or after movement by the pika; in response to calls or movement by a nearby pika; while chasing another pika or while being chased; and when predators are active.

males. The pika killed by a weasel had forayed from his home territory to chase an unfamiliar, immigrant adult male. Juveniles that move away from their natal home range are similarly attacked by residents.

Affiliative behavior is seen in pairs of neighboring males and females, who are not only frequently tolerant of each other but engage in duets of short calls. Such behavior is rarely seen between neighbors of similar sex or between non-neighbor heterosexual pairs.

Adults treat their offspring as they do their neighbors of the opposite sex. Some aggression is directed toward juveniles, but also frequent expressions of social tolerance. Most juveniles remain on the home ranges of their parents throughout the summer.

Ecological constraints have apparently led to a monogamous mating system in pikas. Although males cannot contribute directly to the raising of young (hence they are not monogamous because of their need to assist a single female to raise her young), they still primarily associate with a single neighboring female. Polygyny evolves when males can monopolize sufficient resources to attract several females or when they can directly defend several females. The essentially linear reach of vegetation at the base of the talus precludes resource defense polygyny. Males cannot defend groups of females because they are dispersed and held apart by their mutual antagonism.

Juveniles of both sexes are likely to be repelled should they disperse and attempt to colonize an occupied talus. As a result juveniles normally settle close to their site of birth. This "philopatric" settlement may lead to incestuous matings and contribute to the low genetic variability found in pika populations.

The close association among male-female pairs and the close relatedness of neighbors may underlie the evolution of cooperative behavior patterns in pikas. First, attacks on intruders by residents may be an expression of indirect paternal care: if adults can successfully repel immigrants they may increase the probability of settlement of their offspring should a local site become available for colonization. Second—returning to the opening account—the alarm calls given by both sexes when the weasel struck the resident pika served to warn close kin—note that the unrelated immigrant was the only pika that did not call. Uncontested, the newcomer immediately moved across the talus to claim the slain pika's territory, half-completed haypile, and access to a neighboring female.   ATS

a winning ticket; territories are almost always claimed by a member of the same sex as the previous occupant.

The behavior pattern that may sustain this pattern of occupancy based on sex is apparently a compromise between aggressive and affiliative tendencies. Although all pikas are pugnacious when defending territories, females are less aggressive to neighboring males and more aggressive to females. Male residents rarely exhibit aggression toward each other, simply because they rarely come into contact, but they apparently avoid each other by using scent marking and vocalizations. Males, however, vigorously attack unfamiliar (immigrant)

ELEPHANT-SHREWS

ORDER: MACROSCELIDEA

Fifteen species in 2 subfamilies and 4 genera.
Family: Macroscelididae.
Distribution: N Africa, E, C and S Africa.

Habitat: varies considerably, including montane and lowland forest, savanna, steppe, desert.

Size: varies from the Short-eared elephant-shrew with head-body length 4.1–4.5in (10.4–11.5cm), tail length 4.5–5in (11.5–13cm), weight about 1.6oz (45g) to the Golden-rumped elephant-shrew with a head-body length of 11–12in (27–29.4cm), tail length 9.5–10.5in (23–25.5cm), weight about 19oz (540g).

Gestation: 57–65 days in the Rufous elephant-shrew, about 42 days in the Golden-rumped elephant-shrew.

Longevity: 2½ years in the Rufous elephant-shrew (5½ recorded in captivity), 4 years in the Golden-rumped elephant-shrew.

Species: **Golden-rumped elephant-shrew** (*Rhynchocyon chrysopygus*), **Black and rufous elephant-shrew** (*R. petersi*), **Chequered elephant-shrew** (*R. cirnei*), **Short-snouted elephant-shrew** (*Elephantulus brachyrhynchus*), **Cape elephant-shrew** (*E. edwardi*), **Dusky-footed elephant-shrew** (*E. fuscipes*), **Dusky elephant-shrew** (*E. fuscus*), **Bushveld elephant-shrew** (*E. intufi*), **Eastern rock elephant-shrew** (*E. myurus*), **Somali elephant-shrew** (*E. revoili*), **North African elephant-shrew** (*E. rozeti*), **Rufous elephant-shrew** (*E. rufescens*), **Western rock elephant-shrew** (*E. rupestris*), **Short-eared elephant-shrew** (*Macroscelides proboscideus*), **Four-toed elephant-shrew** (*Petrodromus tetradactylus*).

A NYONE unfamiliar with elephant-shrews might assume that they are large versions of true shrews—the small gray mammals with little beady eyes, pointed snouts and short legs that barely lift their bellies from the ground. Armchair naturalists, never having seen elephant-shrews in the wild, referred to them as jumping shrews because they thought that their long rear legs were used for hopping. Field naturalists in Africa called them elephant-shrews because they have long snouts like elephants and eat invertebrates like shrews. Although elephant-shrews do have long snouts to forage for invertebrates, their similarity to true shrews ends there. But with large eyes and long legs resembling those of small antelope, a trunk-like nose, high-crowned cheek teeth similar to those of a herbivore, and a long rat-like tail they have been shuffled from one taxonomic group to another. At times they have been included in the insectivores (insect-eaters), classed as a type of ungulate, the Menotyphla, which once also included the tree shrews, and placed in their own order, the Macroscelidea. Most systematics now agree that elephant-shrews are indeed unique and belong in their own order, but what is therefore their evolutionary relationship with other mammals?

Recently discovered fossil material and a reinterpretation of dental and foot morphology suggest that rabbits and hares (Lagomorpha) and elephant-shrews had a common Asian ancestor in the Cretaceous era, about 100 million years ago. The elephant-shrews became isolated in Africa, and by the late Oligocene (about 30 million years ago) they occurred in several diverse forms that included small insectivorous forms (Macroscelidinae), small herbivorous species (Mylomygalinae), weighing about 1.8oz (50g) and resembling grass-eating rodents, and large plant-eaters (Myohyracinae), weighing about 18oz (500g) that were so ungulate-like that they were initially misidentified as hyraxes. Today all that remains of these ancient groups are two well-defined insectivorous subfamilies, both still restricted to Africa: the giant elephant-shrews (Rhynchocyoninae) and the small elephant-shrews (Macroscelidinae). The other subfamilies mysteriously died out by the Pleistocene (2 million years ago).

Elephant-shrews are widespread in Africa, occupying habitats as diverse as the Namib Desert in southwest Africa, the steppes and savannas of East Africa, the mountain and lowland forests of central Africa, and semiarid habitats of extreme northwestern Africa. Their absence from west Africa has never been adequately explained. Nowhere are elephant-shrews particularly common, and despite being active above ground during the day and in the evening they are difficult to see. The small species have the size of a mouse and are cryptic in behavior, the larger and more colorful giant elephant-shrews are usually only heard as they bound noisily away into the forest. Both are very secretive.

All species are strictly terrestrial. Despite the diversity of habitats and the difference in size between the smallest and largest species, there is little variation in social organization. Individuals of the Golden-rumped, Four-toed, Short-eared, Rufous and Western rock elephant-shrews live as loosely associated monogamous pairs on contiguous home ranges that are defended against neighboring pairs. As in most mono-

▶ **Representative species of elephant shrews.** (1) Rufous elephant-shrew (*Elephantulus rufescens*) foraging for insects. (2) Chequered elephant-shrew (*Rhynchocyon cirnei*) scent marking with anal glands. (3) Short-eared elephant shrew (*Macroscelides proboscideus*) clearing trail. (4) North African elephant-shrew (*E. rozeti*) face washing at burrow entrance. (5) Four-toed elephant shrew (*Petrodromus tetradactylus*) extruding tongue after insects. (6) Black and rufous elephant shrew (*R. petersi*) tearing prey with teeth and claws. (7) Golden-rumped elephant-shrew (*R. chrysopygus*) stalking before chase. (8) Tail of Four-toed elephant-shrew showing the knobbed bristles uniquely found along the bottom of the tail of some races. One of the earliest suggestions for the occurrence was that they result from scorching in the frequent brush fires. Another idea was that the bristled tail is used as a broom to sweep clean its trails. More recently it has been proposed that they are used to detect ground vibrations, such as other foot-drumming elephant shrews and approaching predators. During aggressive and sexual encounters, this species depresses its tail to the ground and lashes it from side to side, dragging the bristles across the substrate. Perhaps the animals are scent-marking during these encounters, and the knobs act as swabs to spread scent-bearing sebum from the large glands at the base of each bristle.

gamous mammals the sexes are similar in size and appearance, with the exception of the larger canine teeth of the male giant elephant-shrews.

In territorial encounters, visual signals are important but all forms have scent glands that are used to mark their home ranges. These are located on the bottom of the tail in several species of the genus *Elephantulus*, on the soles of the feet in the Rufous elephant-shrew, on the chest of the Dusky-footed, Rufous and Somali elephant-shrews, and just behind the anus in the giant elephant-shrews. Vocal communication is unimportant, though sounds are created by the Four-toed elephant-shrew and some species of *Elephantulus* by drumming their rear feet; *Rhynchocyon* species slap their tails on the ground. When captured the giant elephant-shrews emit a sharp, highpitched scream. When handled they are surprisingly gentle, rarely attempting to bite despite having well-developed canine teeth.

Most species breed year-round. With a gestation of about 45 days for the giant forms and about 60 days for the smaller species; several litters per year are usually produced. Litters normally contain one or two young, born in a well-developed state with a coat pattern similar to that of adults. The North African elephant-shrew and Chequered elephant-shrew may produce three young per litter. The young of giant elephant-shrews require more care than those of the smaller species. They are confined to the nest for several days before they can accompany their mother.

The Four-toed elephant-shrew, Rufous elephant-shrew and Short-eared elephant-shrew clear and maintain complex trail networks to enable them to traverse their territories easily and quickly. Other species may create trails in habitats where vegetation and surface litter is particularly dense. The Short-eared, Western rock and Bushveld elephant-shrews dig short, shallow burrows in sandy substrates, but where the ground is too hard they use abandoned rodent burrows. The Eastern rock elephant-shrew is restricted to rocky areas where it shelters among cracks and crevasses. The most unusual sheltering habits are found in the Four-toed and Rufous elephant-shrews, which use neither burrow nor shelter but spend their entire lives relatively exposed on their trail systems, much as small antelopes do. Their distinct black and white facial pattern probably serves to disrupt the contour of their large black eye, thus camouflaging them from predators. The giant elephant-shrews are more typical of small mammals, in that they spend each night in a leaf nest on the forest floor.

Elephant-shrews spend much of their active hours feeding on invertebrates, and the small species also eat plant matter, especially small fleshy fruits and seeds. The giant forms search for their prey as small coatis or pigs do, using their proboscis-like noses to probe in thick leaf litter and the long claws on their forefeet to excavate in the soil. The small elephant-shrews normally feed by gleaning small invertebrates from the surface of the soil, leaves and twigs. All species have long tongues which will extend to the tips of their noses and are used to flick small items of prey into their mouths.

Elephant-shrews are of little economic importance to man, though along the coast of Kenya the Golden-rumped and Four-toed elephant-shrews are snared and eaten. Recently the Rufous elephant-shrew has been successfully exhibited and bred in numerous zoological gardens, especially in the United

The Trail System of the Rufous Elephant-shrew

Rufous elephant-shrews inhabiting the densely wooded savannas of Tsavo in Kenya are distributed as male-female pairs on territories that vary in size from 17,200–48,400 sq ft (about 1,600 to 4,500 sq m). Although monogamous, individuals spend little time together. Members of a territorial pair cooperate to the extent that they share common boundaries, but when defending their area they behave as individuals, with females only showing aggression towards other females and males towards males. This system of monogamy, characterized by limited cooperation between the sexes, is also found in several small antelopes, such as the dikdik and klipspringer.

There are also similarities between elephant-shrews and some small ungulates in predator avoidance. Camouflage is important in eluding initial detection, but if this fails elephant-shrews use their long legs to flee, swiftly, and outdistance pursuing predators. But how can a 2oz (58g) elephant-shrew, that stands only 2.4in (6cm) high at the shoulder manage to escape the numerous birds of prey, snakes, mongooses, and cats that also inhabit the Tsavo woodlands?

The answer lies partly in the use of a complex network of crisscrossing trails, which each pair builds, maintains and defends. The trails allow the elephant-shrews to take full advantage of their running abilities, providing they are kept immaculately clean. Just a single twig could break an elephant-shrew's flight, with disastrous consequences. Every day individuals of a pair separately traverse much of their trail network, removing accumulated leaves and twigs with swift side-strikes of the forefeet. Paths that are used infrequently are composed of a series of small bare oval patches on the sandy soil, on which an animal lands as it bounds along the trail. Paths that are heavily used become continuous bare channels through the litter.

The Rufous elephant-shrew produces highly precocial and independent young. Since only the female can nurse the young a male can do little to assist directly. Why then is the elephant-shrew monogamous? Part of the answer relates to their system of paths. Males spend nearly twice as much time trail-cleaning as females. Although this indirect help is not as dramatic as the direct cooperation of male wolves and marmosets, it is just as vital to the elephant-shrew's reproductive success, since without paths its ungulate-like habits would be completely ineffective.

States. Giant elephant-shrews are exceedingly difficult to keep in captivity and they have never been bred. Most elephant-shrews are fairly widespread and occupy habitats that have little or no agricultural potential. There is little immediate danger of their numbers being reduced by habitat destruction. The forest-dwelling giant elephant-shrews, however, face severe habitat depletion. This is especially true for the Golden-rumped and Black and rufous elephant-shrews, which occupy small isolated patches of forest that are quickly being cleared for subsistence farming, exotic tree plantations and tourist developments. It would be an incredible loss if these unique, colorful mammals were to disappear, after more than 30 million years of evolution in Africa, just because a few square miles of forest habitat could not be preserved.

GBR

▲ **Aggressive encounters.** Rufous elephant-shrews visibly mark their territories by creating small piles of dung in areas where the paths of two adjoining pairs meet. Occasionally aggressive encounters occur in these territorial arenas. Two animals of the same sex face one another and while slowly walking in opposite directions they stand high on their long legs and accentuate their white feet, much like small mechanical toys. If one of the animals does not retreat, a fight usually develops and the loser is routed from the area.

▶ **Long nose, big eyes** and small size leave little doubt as to why elephant-shrews got their name. However, they are not shrews but separated in a group of their own. Shown here is the Rufous elephant shrew which inhabits wooded savannas of East Africa.

Escape and Protection
The tactics and adaptations of the Golden-rumped elephant-shrew

The sun was just setting when a Golden-rumped elephant-shrew approached an indistinct pile of leaves, about 3ft (1m) wide on the forest floor. The animal paused at the edge of the low mound for 15 seconds, sniffing, listening and watching for the least irregularity. When nothing unusual was sensed, it quietly slipped under the leaves. The leaf nest shuddered for a few seconds as the elephant-shrew arranged itself for the night, then everything was still.

At about the same time the animal's mate was retreating for the night into a similar nest located on the other side of their home range. As this elephant-shrew prepared to enter its nest, a twig snapped. The animal froze, and then quietly left the area for a third nest, which it eventually entered, but not before dusk had fallen.

Every evening, within a few minutes of sunset, these pairs of elephant-shrews separately approach and cautiously enter one of a dozen or more nests they have constructed throughout their home range. Each evening a different nest is used to discourage forest predators such as leopards and eagle-owls from learning that elephant-shrews are always found in a nest.

This is just one of several ways by which Golden-rumped elephant-shrews have learnt to avoid predators. The problem they face is considerable. During the day they spend over 75 percent of their time exposed while foraging in leaf litter on the forest floor. They are the prey of Black mambas, Forest cobras and harrier eagles. To prevent capture by such enemies the Golden-rumped elephant-shrew has developed tactics that involve not only its ability to run fast but also its distinct coat pattern and flashy coloration.

Golden-rumped elephant-shrews can bound across open forest floor at speeds above 16 miles per hour (25 kilometres per hour)—about as fast as an average person can run. Because they are relatively small, they also can pass easily through patches of undergrowth, leaving behind larger terrestrial and aerial predators. Despite speed and agility, however, they are still vulnerable to ambush by sit-and-wait predators, such as the Southern banded harrier eagle. Most small terrestrial mammals have coats or skins with cryptic colors, acting as camouflage. However, the forest floor along the coast of Kenya where the Golden-rumped

▲ **Foraging elephant-shrew.** The Golden-rumped elephant-shrew has a small mouth located far behind the top of its snout, which makes it difficult to ingest large prey items. Small invertebrates are eaten by flicking them into the mouth with a long extensible tongue.

◄ **Daily activities** of the Golden-rumped elephant-shrew. (**a**) Nest-building occurs mainly in the early morning hours, when dead leaves are moist with dew and make little noise. Predators are less likely to be attracted by nest-building activity in the morning. Weathered nests are nearly indistinguishable from the surrounding forest floor. The elephant-shrews use a sleeping posture, with their heads tucked back under their chest. (**b**) The elephant-shrews in the Arabuko-Sokoke forest of coastal Kenya feed mainly on beetles, centipedes, termites, cockroaches, ants, spiders and earthworms, in decreasing order of importance. (**c**) Elephant-shrews chase intruders from their territory using a halfbound gait.

elephant-shrew lives is relatively open so camouflage would be ineffective.

This elephant-shrew's tactic is to "invite" predators to take notice. It has a rump patch that is so visible that a waiting predator will discover a foraging elephant-shrew while it is too far away for making successful ambush. The initial action of a predator detecting an elephant-shrew, such as rapidly turning its head or shifting its weight from one leg to another, may result in enough motion or sound to reveal its presence. By inducing a predator to disclose prematurely its presence or intent to attack, a surprise ambush can be averted. An elephant-shrew that has discovered a predator outside its flight distance does not bound away but pauses and then, with its tail, slaps the leaf litter every few seconds. The sharp sound produced by this behavior probably communicates a message to the predator: "I know you are there, but you are outside my

flight distance, and I can probably outrun you if you attack." Through experience the predator has probably learned that when it hears this signal it is probably futile to attempt a pursuit because the animal is on guard and can easily escape. But what happens, however, if an elephant-shrew unwittingly forages under, for example, an eagle that is perched within its flight distance? As the bird swoops to make its kill the elephant-shrew takes flight across the forest floor towards the nearest cover, noisily pounding the leaf litter with its rear legs as it bounds away. Only speed and agility will save it in this situation.

The Golden-rumped elephant-shrew is monogamous, but pairs spend only about 20 percent of their time in visual contact with each other. The rest they spend resting or foraging alone. So for most of the time they must communicate with scent or sound. The distinct sound of an elephant-shrew tail-slapping or bounding across the forest floot can be heard over a large area of a pair's 3.7 acres (1.5ha) territory. These sounds not only signal to a predator that it has been discovered, but also communicate to an elephant-shrew's mate and young that an intruder has been detected.

Each pair of elephant-shrews defends its territorial boundaries against neighbors and wandering subadults in search of their own territories. During an aggressive encounter a resident will pursue an intruder in a high-speed chase through the forest. If the intruder is not fast enough it will be gashed by the long canines of the resident. These attacks between elephant-shrews might be thought of as a special type of predator–prey interaction, and reveal yet another way in which this animal's coloration may serve to avoid successful predation. The skin under the animal's rump patch is up to three times thicker than the skin on the middle of the back. The golden color of the rump probably serves as a target to discourage attacks on such vital parts of the body as the head and flanks. The toughness of the rump makes it best suited to take attacks. Deflective marks are common in many invertebrates and have been shown to be effective in foiling predators. The distinct eye spots on the wings of some butterflies attract the predatory attacks of birds, allowing the insects to escape relatively unharmed. The yellow rump (and the white tip on the black tail) of the elephant-shrew may serve a similar function by also attracting the talons of an eagle or a striking snake to the rump, thus improving the chances of making a successful escape.   GBR

BIBLIOGRAPHY

The following list of titles indicates key reference works used in the preparation of this volume and those recommended for further reading.

General

Boyle, C. L. (ed) (1981) *The RSPCA Book of British Mammals*, Collins, London.

Corbet, G. B. and Hill, J. E. (1980) *A World List of Mammalian Species*, British Museum and Cornell University Press, London and Ithaca, N.Y.

Grzimek, B. (ed) (1972) *Grzimek's Animal Life Encyclopedia*, vols 10, 11 and 12, Van Nostrand Reinhold, New York.

Hall, E. R. and Kelson, K. R. (1959) *The Mammals of North America*, Ronald Press, New York.

Harrison Matthews, L. (1969) *The Life of Mammals*, vols 1 and 2, Weidenfeld & Nicolson, London.

Honacki, J. H., Kinman, K. E. and Koeppl, J. W. (eds) (1982) *Mammal Species of the World*, Allen Press and Association of Systematics Collections, Lawrence, Kansas.

Kingdon, J. (1971–82) *East African Mammals*, vols I–III, Academic Press, New York.

Morris, D. (1965) *The Mammals*, Hodder & Stoughton, London.

Nowak, R. M. and Paradiso, J. L. (eds) (1983) *Walker's Mammals of the World* (4th edn), 2 vols, Johns Hopkins University Press, Baltimore and London.

Vaughan, T. L. (1972) *Mammalogy*, W. B. Saunders, London and Philadelphia.

Young, J. Z. (1975) *The Life of Mammals: their Anatomy and Physiology*, Oxford University Press, Oxford.

Rodents and Lagomorphs

Barnett, S. A. (1975) *The Rat: A Study in Behavior*, University of Chicago Press, Chicago and London.

Berry, R. J. (ed) (1981) *Biology of the House Mouse*, Academic Press, London.

Calhoun, J. B. (1962) *The Ecology and Welfare of the Norway Rat*, US Public Health Service, Baltimore.

Curry-Lindahl, K. (1980) *Der Berglemming*, A. Ziemsen Verlag, Wittenberg.

Delany, M. J. (1975) *The Rodents of Uganda*, British Museum (Natural History), London.

Eisenberg, J. F. (1963) *The Behavior of Heteromyid Rodents*, University of California Publications in Zoology, vol 69, pp1–100.

Ellerman, J. R. (1940) *The Families and Genera of Living Rodents*. British Museum (Natural History), London.

Ellerman, J. R. (1961) *The Fauna of India: Mammalia*, vol 3, Delhi.

Elton, C. (1942) *Voles, Mice and Lemmings*, Oxford University Press, Oxford.

Errington, P. L. (1963) *Muskrat Populations*, University of Iowa Press, Iowa City.

de Graff, G. (1981) *The Rodents of Southern Africa*, Butterworth, Durban.

Harrison, D. L. (1972) *The Mammals of Arabia*, vol 3. Ernest Benn, London.

King, J. (ed) (1968) *The Biology of Peromyscus*, Special Publication no. 2, American Society of Mammalogists, Oswego, N.Y.

Kingdon, J. (1971) *East African Mammals: An Atlas of Evolution*, vol 2B, Academic Press, London and New York.

Laidler, K. (1980) *Squirrels in Britain*, David and Charles, Newton Abbot and North Pomfret, Vermont.

Linsdale, J. M. (1946) *The California Ground Squirrel*, University of California Press, Berkeley.

Linsdale, J. M. and Tevis, L. P. (1951) *The Dusky-footed Woodrat*, University of California Press, Berkeley.

Lockley, R. M. (1976) *The Private Life of the Rabbit* (2nd edn), André Deutsch, London.

Menzies, J. I. and Dennis, E. (1979) *Handbook of New Guinea Rodents*, Handbook no. 6, Wau Ecology Institute, Wau, New Guinea.

Morgan, L. H. (1868) *The American Beaver and his Works*, Burt Franklin, New York.

Niethammer, J. and Krapp, F. (1978, 1982) *Handbuch der Säugetiere Europas*, vols 1 and 2, Akademische Verlagsgesellschaft, Wiesbaden.

Ognev, S. I. (1963) *Mammals of the USSR and Adjacent Countries*, vols 4 and 6, Israel Program for Scientific Translation, Jerusalem.

Orr, R. T. (1977) *The Little-known Pika*, Collier Macmillan, New York.

Prakash, I. and Ghosh, P. K. (eds) (1975) *Rodents in Desert Environments*, Monographae Biologicae, W. Junk, The Hague.

Rosevear, D. R. (1969) *The Rodents of West Africa*, British Museum (Natural History), London.

Rowlands, I. W. and Weir, B. (1974) *The Biology of Hystricomorph Rodents*, Zoological Society Symposia, no. 34. Academic Press, London and New York.

Watts, C. H. S. and Aslin, H. J. (1981) *The Rodents of Australia*, Angus and Robertson, Sydney and London.

GLOSSARY

Adaptive radiation the pattern in which different species develop from a common ancestor (as distinct from CONVERGENT EVOLUTION, a process whereby species from different origins became similar in response to the same SELECTIVE PRESSURES).

Adult a fully developed and mature individual capable of breeding, but not necessarily doing so until social and/or ecological conditions allow.

Aestivate (noun: aestivation) to enter a state of dormancy or torpor in seasonal hot, dry weather, when food is scarce.

Allantoic stalk a sac-like outgrowth of the hinder part of the gut of the mammalian fetus, containing a rich network of blood vessels. It connects fetal circulation with the PLACENTA, facilitating nutrition of the young, respiration and excretion. (See CHORIOALLANTOIC PLACENTATION.)

Allopatry condition in which populations of different species are geographically separated (cf SYMPATRY).

Alpine of the Alps or any lofty mountains; usually pertaining to altitudes above 4,900ft (1,500m).

Altricial young that are born at a rudimentary stage of development and require an extended period of nursing by parent(s). See also PRECOCIAL.

Amphibious able to live on both land and in water.

Anal gland or sac a gland opening by a short duct either just inside the anus or on either side of it.

Ancestral stock a group of animals, usually showing primitive characteristics, which is believed to have given rise to later, more specialized forms.

Antigen a substance, whether organic or inorganic, that stimulates the production of antibodies when introduced into the body.

Antrum a cavity in the body, especially one in the upper jaw bone.

Aquatic living chiefly in water.

Arboreal living in trees.

Astragalus a bone in the ungulate tarsus (ankle) which (due to reorganization of ankle bones following reduction in the number of digits) bears most of the body weight (a task shared by the CALCANEUM bone in most other mammals).

Baculum (os penis or penis bone) an elongate bone present in the penis of certain mammals.

Bifid (of the penis) the head divided into two parts by a deep cleft.

Biotic community a naturally occurring group of plants and animals in the same environment.

Blastocyst see IMPLANTATION.

Boreal region a zone geographically situated south of the Arctic and north of latitude 50°N; dominated by coniferous forest.

Brachydont a type of short-crowned teeth whose growth ceases when full-grown, whereupon the pulp cavity in the root closes. Typical of most mammals, but contrast the HYPSODONT teeth of many herbivores.

Brindled having inconspicuous dark streaks or flecks on a gray or tawny background.

Browser a herbivore which feeds on shoots and leaves of trees, shrubs etc, as distinct from grasses (cf GRAZER).

Bullae (auditory) globular, bony capsules housing the middle and inner ear structures, situated on the underside of the skull.

Bunodont molar teeth whose cusps form separate, rounded hillocks which crush and grind.

Cache a hidden store of food; also (verb) to hide food for future use.

Calcaneum one of the tarsal (ankle) bones which forms the heel and in many mammalian orders bears the body weight together with the ASTRAGALUS.

Cannon bone a bone formed by the fusion of METATARSAL bones in the feet of some families.

Carnivore any meat-eating organism (alternatively, a member of the order Carnivora, many of whose members are carnivores).

Carpals wrist bones which articulate between the forelimb bones (radius and ulna) and the METACARPALS.

Caudal gland an enlarged skin gland associated with the root of the tail. Subcaudal: placed below the root; supracaudal: above the root.

Cecum a blind sac in the digestive tract, opening out from the junction between the small and large intestines. In herbivorous mammals it is often very large; it is the site of bacterial action on cellulose. The end of the cecum is the appendix; in species with reduced ceca the appendix may retain an antibacterial function.

Cellulose the fundamental constituent of the cell walls of all green plants, and some algae and fungi. It is very tough and fibrous, and can be digested only by the intestinal flora in mammalian guts.

Cementum hard material which coats the roots of mammalian teeth. In some species, cementum is laid down in annual layers which, under a microscope, can be counted to estimate the age of individuals.

Cervix the neck of the womb (cervical—pertaining to neck).

Cheek-teeth teeth lying behind the canines in mammals, comprising premolars and molars.

Chorioallantoic placentation a system whereby fetal mammals are nourished by the blood supply of the mother. The chorion is a superficial layer enclosing all the embryonic structures of the fetus, and is in close contact with the maternal blood supply at the placenta. The union of the chorion (with its vascularized ALLANTOIC STALK and YOLK SAC) with the placenta facilitates the exchange of food substances and gases, and hence the nutrition of the growing fetus.

Class taxonomic category subordinate to a phylum and superior to an order.

Clavicle the collar bone.

Cloaca terminal part of the gut into which the reproductive and urinary ducts open. There is one opening to the body, the cloacal aperture, instead of a separate anus and urinogenital opening.

Cloud forest moist, high-altitude forest characterized by dense UNDERSTORY growth, and abundance of ferns, mosses, orchids and other plants on the trunks and branches of the trees.

Colon the large intestine of vertebrates, excluding the terminal rectum. It is concerned with the absorption of water from feces.

Colonial living together in colonies. In bats, more usually applied to the communal sleeping habit, in which tens of thousands of individuals may participate.

Concentrate selector a herbivore which

feeds on those plant parts (such as shoots and fruits) which are rich in nutrients.

Congenor a member of the same species (or genus).

Conspecific member of the same species.

Convergent evolution the independent acquisition of similar characters in evolution, as opposed to possession of similarities by virtue of descent from a common ancestor.

Crepuscular active in twilight.

Cricetine adjective and noun used to refer to (a) the primitive rodents from which the New World rats and mice, voles and lemmings, hamsters and gerbils are descended, (b) these modern rodents. In some taxonomic classification systems these subfamilies of the family Muridae are classified as members of a separate family called Cricetidae, with members of the Old World rats and mice alone constituting the Muridae.

Crypsis an aspect of the appearance of an organism which camouflages it from the view of others, such as predators or competitors.

Cryptic (coloration or locomotion) protecting through concealment.

Cue a signal, or stimulus (eg olfactory) produced by an individual which elicits a response in other individuals.

Cursorial being adapted for running.

Cusp a prominence on a cheek-tooth (premolar or molar).

Delayed implantation see IMPLANTATION.

Dental formula a convention for summarizing the dental arrangement whereby the numbers of each type of tooth in each half of the upper and lower jaw are given; the numbers are always presented in the order: incisor (I), canine (C), premolar (P), molar (M). The final figure is the total number of teeth to be found in the skull. A typical example for Carnivora would be I3/3, C1/1, P4/4, M3/3 = 44.

Dentition the arrangement of teeth characteristic of a particular species.

Dermis the layer of skin lying beneath the outer epidermis.

Desert areas of low rainfall, typically with sparse scrub or grassland vegetation or lacking vegetation altogether.

Dicotyledon one of the two classes of flowering plants (the other class comprises monocotyledons), characterized by the presence of two seed leaves in the young plant, and by net-veined, often broad leaves, in mature plants. Includes deciduous trees, roses etc.

Digit a finger or toe.

Digital glands glands that occur between or on the toes.

Digitigrade method of walking on the toes without the heel touching the ground (cf plantigrade).

Disjunct or **discontinuous distribution** geographical distribution of a species that is marked by gaps. Commonly brought about by fragmentation of suitable habitat, especially as a result of human intervention.

Dispersal the movements of animals, often as they reach maturity, away from their previous home range (equivalent to emigration). Distinct from dispersion, that is, the pattern in which things (perhaps animals, food supplies, nest sites) are distributed or scattered.

Display any relatively conspicuous pattern of behaviour that conveys specific information to others, usually to members of the same species; can involve visual and/or vocal elements, as in threat, courtship or "greeting" displays.

Distal far from the point of attachment or origin (eg tip of tail).

Diurnal active in daytime.

Dormancy a period of inactivity; many bears, for example, are dormant for a period in winter; this is not true HIBERNATION, as pulse rate and body temperature do not drop markedly.

Dorsal on the upper or top side or surface (eg dorsal stripe).

Ecology the study of plants and animals in relation to their natural environmental setting. Each species may be said to occupy a distinctive ecological NICHE.

Ecosystem a unit of the environment within which living and nonliving elements interact.

Ecotype a genetic variety within a single species, adapted for local ecological conditions.

Edentate a member of an order comprising living and extinct anteaters, sloths, armadillos (XENARTHRANS), and extinct paleanodonts.

Elongate relatively long (eg of canine teeth, longer than those of an ancestor, a related animal, or than adjacent teeth).

Embryonic diapause the temporary cessation of development of an embryo (eg in some bats and kangaroos).

Emigration departure of animal(s), usually at or about the time of reaching adulthood, from the group or place of birth.

Epidermis the outer layer of mammalian skin (and in plants the outer tissue of young stem, leaf or root).

Erectile capable of being raised to an erect position (erectile mane).

Estrus the period in the estrous cycle of female mammals at which they are often attractive to males and receptive to mating. The period coincides with the maturation of eggs and ovulation (the release of mature eggs from the ovaries). Animals in estrus are often said to be "on heat" or "in heat." In primates, if the egg is not fertilized the subsequent degeneration of uterine walls (endometrium) leads to menstrual bleeding. In some species ovulation is triggered by copulation and this is called **induced ovulation**, as distinct from spontaneous ovulation.

Eucalypt forest Australian forest, dominated by trees of the genus *Eucalyptus*.

Eutherian a mammal of the subclass Eutheria, the dominant group of mammals. The embryonic young are nourished by an ALLANTOIC PLACENTA.

Exudate natural plant exudates include gums and resins; damage to plants (eg by marmosets) can lead to loss of sap as well.

Facultative optional (cf. OBLIGATE).

Family a taxonomic division subordinate to an order and superior to a genus.

Feces excrement from the bowels; colloquially known as droppings or scats.

Feral living in the wild (of domesticated animals, eg cat, dog).

Fermentation the decomposition of organic substances by microorganisms. In some mammals, parts of the digestive tract (eg the cecum) may be inhabited by bacteria that break down cellulose and release nutrients.

Fetal development rate the rate of development, or growth, of unborn young.

Flehmen German word describing a facial expression in which the lips are pulled back, head often lifted, teeth sometimes clapped rapidly together and nose wrinkled. Often associated with animals (especially males) sniffing scent marks or socially important odors (eg scent of estrous female). Possibly involved in transmission of odor to JACOBSON'S ORGAN.

Folivore an animal eating mainly leaves.

Follicle a small sac, therefore (a) a mass of ovarian cells that produces an ovum (b) an indentation in the skin from which hair grows.

Forbs a general term applied to ephemeral or weedy plant species (not grasses). In arid and semi-arid regions they grow abundantly and profusely after rains.

Fossorial burrowing (of life-style or behavior).

Frugivore an animal eating mainly fruits.

Gallery forest luxuriant forest lining the banks of watercourses.

Gamete a male or female reproductive cell (ovum or spermatozoon).

Generalist an animal whose life-style does not involve highly specialized stratagems (cf SPECIALIST); for example, feeding on a variety of foods which may require different foraging techniques.

Genotype the genetic constitution of an organism, determining all aspects of its appearance, structure and function.

Genus (plural genera) a taxonomic division superior to species and subordinate to family.

Gestation the period of development within the uterus; the process of **delayed implantation** can result in the period of pregnancy being longer than the period during which the embryo is actually developing (See also IMPLANTATION.)

Glands (marking) specialized glandular areas of the skin, used in depositing SCENT MARKS.

Graviportal animals in which the weight is carried by the limbs acting as rigid, extensible struts, powered by extrinsic muscles; eg elephants and rhinos.

Grazer a herbivore which feeds upon grasses (cf BROWSER).

Grizzled sprinkled or streaked with gray.

Gumivorous feeding on gums (plant exudates).

Harem group a social group consisting of a single adult male, at least two adult females and immature animals; a common pattern of social organization among mammals.

Heath low-growing shrubs with woody stems and narrow leaves (eg heather), which often predominate on acidic or upland soils.

Herbivore an animal eating mainly plants or parts of plants.

Heterothermy a condition in which the internal temperature of the body follows the temperature of the outside environment.

Hibernation a period of winter inactivity during which the normal physiological process is greatly reduced and thus during which the energy requirements of the animal are lowered.

Hindgut fermenter herbivores among which the bacterial breakdown of plant tissue occurs in the CECUM, rather than in the RUMEN or foregut.

Holarctic realm a region of the world including North America, Greenland, Europe, and Asia apart from the southwest, southeast and India.

Home range the area in which an animal normally lives (generally excluding rare excursions or migrations), irrespective of whether or not the area is defended from other animals (cf TERRITORY).

Hybrid the offspring of parents of different species.

Hypothermy a condition in which internal body temperature is below normal.

Hypsodont high-crowned teeth, which continue to grow when full-size and whose pulp cavity remains open; typical of herbivorous mammals (cf BRACHYDONT).

Implantation the process whereby the free-floating blastocyst (early embryo) becomes attached to the uterine wall in mammals. At the point of implantation a complex network of blood vessels develops to link mother and embryo (the placenta). In **delayed implantation** the blastocyst remains dormant in the uterus for periods varying, between species from 12 days to 11 months. Delayed implantation may be obligatory or facultative and is known for some members of the Carnivora and Pinnipedia and others.

Induced ovulation see ESTRUS.

Inguinal pertaining to the groin.

Insectivore an animal eating mainly arthropods (insects, spiders).

Interdigital pertaining to between the digits.

Interfemoral a membrane stretching between the femora, or thigh bones in bats.

Intestinal flora simple plants (eg bacteria) which live in the intestines, especially the CECUM, of mammals. They produce enzymes which break down the cellulose in the leaves and stems of green plants and convert it to digestible sugars.

Introduced of a species which has been brought, by man, from lands where it occurs naturally to lands where it has not previously occurred. Some introductions are accidental (eg rats which have traveled unseen on ships), but some are made on purpose for biological control, farming or other economic reasons (eg the Common brush-tail possum, which was introduced to New Zealand from Australia to establish a fur industry).

Invertebrate an animal which lacks a backbone (eg insects, spiders, crustaceans).

Ischial pertaining to the hip.

Jacobson's organ a structure in a foramen (small opening) in the palate of many vertebrates which appears to be involved in olfactory communication. Molecules of scent may be sampled in these organs.

Juvenile no longer possessing the characteristics of an infant, but not yet fully adult.

Kopje a rocky outcrop, typically on otherwise flat plains of African grasslands.

Labile (body temperature) an internal body temperature which may be lowered or raised from an average body temperature.

Lactation (verb: lactate) the secretion of milk, from MAMMARY GLANDS.

Lamoid Llama-like; one of the South American cameloids.

Larynx dilated region of upper part of windpipe, containing vocal chords. Vibration of chords produces vocal sounds.

Latrine a place where feces are regularly left (often together with other SCENT MARKS); associated with olfactory communication.

Lek a display ground at which individuals of one sex maintain miniature territories into which they seek to attract potential mates.

Lipotyphlan an early insectivore classification; menotyphlan insectivores possess a cecum, lipotyphlans do not. Only lipotyphlans are now classified as Insectivora.

Llano South American semi-arid savanna country, eg. of Venezuela.

Loph a transverse ridge on the crown of molar teeth.

Lophodont molar teeth whose cusps form ridges or LOPHS.

Mallee a grassy, open woodland habitat characteristic of many semi-arid parts of Australia. "Mallee" also describes the multi-stemmed habit of eucalypt trees which dominate this habitat.

Mamma (pl. mammae) **mammary glands** the milk-secreting organ of female mammals, probably evolved from sweat glands.

Mammal a member of the CLASS of VERTEBRATE animals having MAMMARY GLANDS which produce milk with which they nurse their young (properly: Mammalia).

Mammilla (pl. mammillae) nipple, or teat, on the MAMMA of female mammals; the conduit through which milk is passed from the mother to the young.

Mandible the lower jaw.

Mangrove forest tropical forest developed on sheltered muddy shores of deltas and estuaries exposed to tide. Vegetation is almost entirely woody.

Marine living in the sea.

Masseter a powerful muscle, subdivided into parts, joining the MANDIBLE to the upper jaw. Used to bring jaws together when chewing.

Melanism darkness of color due to presence of the black pigment melanin.

Menotyphlan see LIPOTYPHLAN

Metabolic rate the rate at which the chemical processes of the body occur.

Metabolism the chemical processes occurring within an organism, including the production of PROTEIN from amino acids, the exchange of gasses in respiration, the liberation of energy from foods and innumerable other chemical reactions.

Metacarpal bones of the hand, between the CARPALS of the wrist and the phalanges of the digits.

Metapodial the proximal element of a digit (contained within the palm or sole). The metapodial bones are METACARPALS in the hand (manus) and METATARSALS in the foot (pes).

Metatarsal bones of the foot articulating between the tarsals of the ankle and the phalanges of the digits.

Microhabitat the particular parts of the habitat that are encountered by an individual in the course of its activities.

Midden a dunghill, or site for the regular deposition of feces by mammals.

Migration movement, usually seasonal, from one region or climate to another for purposes of feeding or breeding.

Monogamy a mating system in which individuals have only one mate per breeding season.

Monotypic a genus comprising a single species.

Montane pertaining to mountainous country.

Montane forest forest occurring at middle altitudes on the slopes of mountains, below the alpine zone but above the lowland forests.

Morphology (morphological) the structure and shape of an organism.

Moss forest moist forest occurring on higher mountain slopes—eg 4,900–10,500ft (1,500–3,200m) in New Guinea—characterized by rich growths of mosses and other plants on the trunks and branches of the trees.

Murine adjective and noun used to refer to members of the subfamily (of the family Muridae) Murinae, which consists of the Old World rats and mice. In some taxonomic classification systems this subfamily is given the status of a family, Muridae, and the members then are sometimes referred to as murids. See also CRICETINE.

Mutation a structural change in a gene which can thus give rise to a new heritable characteristic.

Natal range the home range into which an individual was born (natal = of or from one's birth).

Niche the role of a species within the community, defined in terms of all aspects of its life-style (eg food, competitors, predators, and other resource requirements).

Nocturnal active at nighttime.

Obligate required, binding (cf FACULTATIVE).

Occipital pertaining to the occiput at the back of the head.

Olfaction, olfactory the olfactory sense is the sense of smell, depending on receptors located in the epithelium (surface membrane) lining the nasal cavity.

Omnivore an animal eating a varied diet including both animal and plant tissue.

Opposable (of first digit) of the thumb and forefinger in some mammals, which may be brought together in a grasping action, thus enabling objects to be picked up and held.

Opportunist (of feeding) flexible behavior of exploiting circumstances to take a wider range of food items; characteristic of many species. See GENERALIST; SPECIALIST.

Order a taxonomic division subordinate to class and superior to family.

Ovulation (verb ovulate) the shedding of mature ova (eggs) from the ovaries where they are produced (see ESTRUS).

Pair-bond an association between a male and female, which lasts from courtship at least until mating is completed, and in some species, until the death of one partner.

Palearctic a geographical region encompassing Europe and Asia north of the Himalayas, and Africa north of the Sahara.

Paleothere a member of the family Paleotheriidae (order Perissodactyla), which became extinct in the early Tertiary.

Palmate palm shaped.

Pampas Argentinian steppe grasslands.

Papilla (plural: papillae) a small nipple like projection.

Páramo alpine meadow of northern and western South American uplands.

Parturition the process of giving birth (hence *post partum* —after birth).

Patagium a gliding membrane typically stretching down the sides of the body between the fore- and hindlimbs and perhaps including part of the tail. Found in colugos, flying squirrels, bats etc.

Perineal glands glandular tissue occurring between the anus and genitalia.

Pheromones secretions whose odors act as chemical messengers in animal communication, and which prompt a specific response on behalf of the animal receiving the message (see SCENT MARK).

Phylogeny a classification or relationship based on the closeness of evolutionary descent.

Phylogenetic (of classification or relationship) based on the closeness of evolutionary descent.

Phylum a taxonomic division comprising a number of classes.

Physiology study of the processes which go on in living organisms.

Pinna (plural: pinnae) the projecting cartilaginous portion of the external ear.

Placenta, placental mammals a structure that connects the fetus and the mother's womb to ensure a supply of nutrients to the fetus and removal of its waste products. Only placental mammals have a well-developed placenta; marsupials have a rudimentary placenta or none and monotremes lay eggs.

Polyandrous see POLYGYNOUS.

Polyestrous having two or more ESTRUS cycles in one breeding season.

Polygamous a mating system wherein an individual has more than one mate per breeding season.

Polygynous a mating system in which a male mates with several females during one breeding season (as opposed to polyandrous, where one female mates with several males).

Polymorphism occurrence of more than one MORPHOLOGICAL form of individual in a population. (See SEXUAL DIMORPHISM.)

Population a more or less separate (discrete) group of animals of the same species within a given BIOTIC COMMUNITY.

Post-orbital bar a bony strut behind the eye-socket (orbit) in the skull.

Post-partum estrus ovulation and an increase in the sexual receptivity of female mammals, hours or days after the birth of a litter (see ESTRUS).

Prairie North American steppe grassland between 30°N and 55°N.

Predator an animal which forages for live prey; hence "anti-predator behavior" describes the evasive actions of the prey.

Precocial of young born at a relatively advanced stage of development, requiring a short period of nursing by parents (see ALTRICIAL).

Prehensile capable of grasping.

Pre-orbital in front of the eye socket.

Preputial pertaining to the prepuce or loose skin covering the penis.

Proboscidean a member of the order of primitive ungulates, Proboscidea.

Proboscis a long flexible snout.

Process (anatomical) an outgrowth or protuberance.

Procumbent (incisors) projecting forward more or less horizontally.

Promiscuous a mating system wherein an individual mates more or less indiscriminately.

Protein a complex organic compound made of amino acids. Many different kinds of proteins are present in the muscles and tissues of all mammals.

Proximal near to the point of attachment or origin, (eg the base of the tail).

Puberty the attainment of sexual maturity. In addition to maturation of the primary sex organs (ovaries, testes), primates may exhibit "secondary sexual characteristics" at puberty. Among higher primates it is usual to find a growth spurt at the time of puberty in males and females.

Puna a treeless tableland or basin of the high Andes.

Pylorus the region of the stomach at its intestinal end, which is closed by the pyloric sphincter.

Quadrate bone at rear of skull which serves as a point of articulation for lower jaw.

Quadrupedal walking on all fours, as opposed to walking on two legs (BIPEDAL) or moving suspended beneath branches in trees.

Race a taxonomic division subordinate to subspecies but linking populations with similar distinct characteristics.

Radiation see ADAPTIVE RADIATION.

Rain forest tropical and subtropical forest with abundant and year-round rainfall. Typically species rich and diverse.

Range (geographical) area over which an organism is distributed.

Receptive state of a female mammal ready to mate or in ESTRUS.

Reduced (anatomical) of relatively small dimension (eg of certain bones, by comparison with those of an ancestor or related animals).

Refection process in which food is excreted and then reingested a second time to ensure complete digestion, as in the Common shrew.

Reingestion process in which food is digested twice, to ensure that the maximum amount of energy is extracted from it. Food may be brought up from the stomach to the mouth for further chewing before reingestion, or an individual may eat its own feces (see REFECTION).

Reproductive rate the rate of production of offspring; the net productive rate may be defined as the average number of female offspring produced by each female during her entire lifetime.

Resident a mammal which normally inhabits a defined area, whether this is a HOME RANGE or a TERRITORY.

Retractile (of claws) able to be withdrawn into protective sheaths.

Rodent a member of the order of Rodentia, the largest mammalian order, which includes rats and mice, squirrels, porcupines, capybara etc.

Rut a period of sexual excitement; the mating season.

Satellite male an animal excluded from the core of the social system but loosely associated on the periphery, in the sense of being a "hanger-on" or part of the retinue of more dominant individuals.

Scent gland an organ secreting odorous material with communicative properties; see SCENT MARK.

Scent mark a site where the secretions of scent glands, or urine or FECES, are deposited and which has communicative significance. Often left regularly at traditional sites which are also visually conspicuous. Also the "chemical message" left by this means; and (verb) to leave such a deposit.

Sclerophyll forest a general term for the hard-leafed eucalypt forest that covers much of Australia.

Scute a bony plate, overlaid by horn, which is derived from the outer layers of the skin. In armadillos, bony scute plates provide armor for all the upper, outer surfaces of the body.

Secondary sexual character a characteristic of animals which differs between the two sexes, but excluding the sexual organs and associated structures.

Sedentary pertaining to mammals which occupy relatively small home ranges and exhibit weak dispersal or migratory tendencies.

Selective pressure a factor affecting the reproductive success of individuals (whose success will depend on the fitness, ie the extent to which they are adapted to thrive under that selective pressure).

Selenodont of molar teeth with crescent-shaped cusps.

Sella one of the nasal processes of leafnose bats; an upstanding central projection which may form a fluted ridge running backwards from between the nostrils.

Septum a partition separating two parts of an organism. The nasal septum consists of a fleshy part separating the nostrils and a vertical, bony plate dividing the nasal cavity.

Serum blood from which corpuscles and clotting agents have been removed; a clear, almost colorless fluid.

Sexual dimorphism a condition in which males and females of a species differ consistently in form, eg size, shape. (See POLYMORPHISM.)

Serology the study of blood sera; investigates ANTIGEN-antibody reactions to elucidate responses to disease organisms and also PHYLOGENETIC relationships between species.

Siblings individuals who share one or both parents. An individual's siblings are its brothers and sisters, regardless of their sex.

Sinus a cavity in bone or tissue.

Sirenia an order of herbivorous aquatic mammals, comprising the manatees and dugong.

Solitary living on its own, as opposed to social or group-living in life-style.

Sonar sound used in connection with Navigation (SOund NAvigation Ranging).

Specialist an animal whose life-style involves highly specialized stratagems: eg feeding with one technique on a particular food.

Species a taxonomic division subordinate to genus and superior to subspecies. In general a species is a group of animals which are similar in structure and are able to breed and produce viable offspring.

Speciation the process by which new species arise in evolution. It is widely accepted that it occurs when a single species population is divided by some geographical barrier.

Sphincter a ring of smooth muscle around a pouch, rectum or other hollow organ, which can be contracted to narrow or close the entrance to the organ.

Spinifex a grass which grows in large, distinctive clumps or hummocks in the driest areas of central and Western Australia.

Stridulation production of sound by rubbing together modified surfaces of the body.

Subadult no longer an infant or juvenile but not yet fully adult physically and/or socially.

Subfamily a division of a FAMILY.

Suborder a subdivision of an ORDER.

Subspecies a recognizable subpopulation of a single species, typically with a distinct geographical distribution.

Successional habitat a stage in the progressive change in composition of a community of plants, from the original colonization of a bare area toward a largely stable climax.

Supra-orbital pertaining to above the eye (eye-socket or orbit).

Sympatry a condition in which the geographical ranges of two or more different species overlap (cf ALLOPATRY).

Syndactylous pertaining to the second and third toes of some mammals, which are joined together so that they appear to be a single toe with a split nail (as opposed to didactylous). In kangaroos, these syndactyl toes are used as a fur comb.

Taiga northernmost coniferous forest, with open boggy rocky areas in between.

Tarsal pertaining to the tarsus bones in the ankle, articulating between the tibia and fibia of the leg and metatarsals of the foot (pes).

Terrestrial living on land.

Territory an area defended from intruders by an individual or group. Originally the term was used when ranges were exclusive and obviously defended at their borders. A more general definition of territoriality allows some overlap between neighbors by defining territoriality as a system of spacing wherein home ranges do not overlap randomly—that is, the location of one individual's, or group's home range influences those of others.

Testosterone a male hormone synthesized in the testes and responsible for the expression of many male characteristics (contrast the female hormone estrogen produced in the ovaries).

Thermoneutral range the range in outside environmental temperature in which a mammal uses the minimum amount of energy to maintain a constant internal body temperature. The limits to the thermoneutral range are the lower and upper critical temperatures, at which points the mammals must use increasing amounts of energy to maintain a constant body temperature. (cf HETEROTHERMY).

Thermoregulation the regulation and maintenance of a constant internal body temperature in mammals.

Tooth-comb a dental modification in which the incisor teeth form a comb-like structure.

Thoracic pertaining to the thorax or chest.

Torpor a temporary physiological state in some mammals, akin to short-term hibernation, in which the body temperature drops and the rate of METABOLISM is reduced. Torpor is an adaptation for reducing energy expenditure in periods of extreme cold or food shortage.

Tragus a flap, sometimes moveable, situated in front of the opening of the outer ear in bats.

Trypanosome a group of protozoa causing

sleeping sickness.

Umbilicus navel.

Underfur the thick soft undercoat fur lying beneath the longer and coarser hair (guard hairs).

Understory the layer of shrubs, herbs and small trees beneath the forest canopy.

Vascular of, or with, vessels which conduct blood and other body fluids.

Vector an individual or species which transmits a disease.

Ventral on the lower or bottom side or surface; thus ventral or abdominal glands occur on the underside of the abdomen.

Vertebrate an animal with a backbone; a division of the phylum Chordata which includes animals with notochords (as distinct from invertebrates).

Vestigial a characteristic with little or no contemporary use, but derived from one which was useful and well developed in an ancestral form.

Vibrissae stiff, coarse hairs richly supplied with nerves, found especially around the snout, and with a sensory (tactile) function.

Vocalization calls or sounds produced by the vocal cords of a mammal, and uttered through the mouth. Vocalizations differ with the age and sex of mammals but are usually similar within a species.

Yolk sac a sac, usually containing yolk, which hangs from the ventral surface of the vertebrate fetus. In mammals, the yolk sac contains no yolk, but helps to nourish the embryonic young via a network of blood vessels.

INDEX

A **bold number** indicates a major section of the main text, following a heading; a ***bold italic*** number indicates a fact box on a single species; a single number in (parentheses) indicates that the animal name or subjects are to be found in a boxed feature and a double number in (parentheses) indicates that the animal name or subjects are to be found in a special spread feature. *Italic* numbers refer to illustrations.

Picture Acknowledgments

Key: *t* top. *b* bottom. *c* center. *l* left. *r* right.
Abbreviations: A Ardea. AH Andrew Henley. AN Nature, Agence Photographique. BC Bruce Coleman Ltd. GF George Frame. J Jacana. NHPA Natural History Photographic Agency. OSF Oxford Scientific Films. PEP Planet Earth Pictures. SA Survival Anglia.

Cover OSF/J. Paling. 1 A/W. Weisser. 2–3 BC/P. Ward. 4–5BC/J. Foott. 6–7 BC/A. Purcell. 8–9A/P. Morris. 10/E. Beaton. 11/AH. 14–15/A. 16/AN. 17/ICI Plant Protection Division. 18/Mansell Collection. 18–19/ICI Plant Protection Division. 19/AH. 20–21/A. Bannister. 22–23/Aquila. 23b, 24, 25b/BC. 25t/J. Kaufmann. 27/BC. 28–29, 29, 30, 30–1/A. 31b/K. Sugowara, Orion Press. 34/Aquila. 35b/M. Fogden. 35t, 36–7, 39t/W. Ervin, Natural Imagery. 39b/NHPA. 40, 41/Bio-Tec Images. 42–3/OSF. 43/SA. 46h/A. 46–7/BC. 49/A. 50/NHPA. 51/J. 53/AH. 54–5/AN. 56/R.W. Barbour. 57t/M. Fogden. 57b/BC. 58/M. Fogden. 59, 64b/BC. 64t, 65/M. Fogden. 66–7/NHPA. 67b/BC. 71/NHPA. 72/Bodleian Library. 73t/J. 73b/Naturfotografernas Bildbyrå. 74–5/BC. 78/AH. 79/NHPA. 80t Scala. 80–81/A. 81t/GF. 84–5/BC. 88/NHPA. 89/BC. 90–1, 94/A. 95/BC. 96–7/AN. 98/R. W. Barbour. 99/AN. 100/A. Bannister. 101/BC. 102–3/Tony Morrison. 103b/BC. 104/GF. 105/BC. 107t/T. Owen-Edmunds. 107b/Tony Morrison. 108–9/A. 110–11/BC. 110b/D. Macdonald. 111c/BC. 111b/NHPA. 112, 113, 114, 115t/D. Macdonald. 115b, 116–17/A. 117b/BC. 121t/A. Bannister. 121b/A. 122, 123/W. George. 124–5/A. Bannister. 125t/P. van den Elzen. 127./J. U. M. Jarvis. 129/PEP. 132/AN. 133/M. Fogden. 133b/A. 134–5/A. 138/A. 139t/BC. 139t/NHPA. 140–1, 141b, 142, 143/BC. 144l/W. Ervin, Natural Imagery. 144–5/BC. 144r/W. Ervin, Natural Imagery. 149/BC. 150–1/P. Morris.

Artwork

All artwork © Priscilla Barrett unless stated otherwise below.
Abbreviations: SD Simon Driver. AEM Anne-Elise Martin.

13c, 20/SD. 21/AEM. 24, 26, 38, 40, 52/SD. 53/AEM. 70, 71, 85, 100/SD. 101/AEM. 127, 128, 129, 138, 140/SD. Maps and scale drawings/SD.